Christina Schmidt

Compaction and cementation controls on reservoir quality in Buntsandstein red beds

Compaction and cementation controls on reservoir quality in Buntsandstein red beds

by
Christina Schmidt

KIT Scientific Publishing

Karlsruher Institut für Technologie
Institut für Angewandte Geowissenschaften

Compaction and cementation controls on
reservoir quality in Buntsandstein red beds

Zur Erlangung des akademischen Grades eines Doktors der Naturwissenschaften von der KIT-Fakultät für für Bauingenieur-, Geo- und Umweltwissenschaften des Karlsruher Instituts für Technologie (KIT) genehmigte Dissertation

von Christina Schmidt
Tag der mündlichen Prüfung: 16. Dezember 2020
Erster Gutachter: Prof. Dr. Christoph Hilgers
Zweiter Gutachter: Prof. Dr. Reinhard Gaupp

Impressum

KIT Scientific Publishing

Karlsruher Institut für Technologie (KIT)
KIT Scientific Publishing
Straße am Forum 2
D-76131 Karlsruhe

KIT Scientific Publishing is a registered trademark
of Karlsruhe Institute of Technology.
Reprint using the book cover is not allowed.

www.ksp.kit.edu

Print on Demand 2022 – Gedruckt auf FSC-zertifiziertem Papier

ISBN 978-3-7315-1154-0
DOI 10.5445/KSP/1000140255

Acknowledgements

I would like to thank my advisor, Christoph Hilgers, for his trust and support. He gave me full responsibility for my tasks in both, my PhD project and the tuition exchange with Moroccan and Tunisian universities. Such that I got to leave my comfort zone and identify unknown strengths and abilities.

Thanks also go to Reinhard Gaupp as my second referee. He thoroughly reviewed my work and with his great interest in the bleaching subject improved it by far.

This work would not have been possible, in the truest sense of the word, if not for Paulus Sinner from the Bamberger Natursteinwerke, who enabled me to enter their Röttbach quarry and was eager to share his knowledge and the sedimentological discoveries he made with great regularity. I am also thankful to the Göhrig Steinmetzbetrieb & Restaurierungen GmbH, which allowed me to enter their production site, the quarry Lahr-Kuhbach, and allowed sampling and supported me on-site.

I am grateful to my colleagues for their support, first and foremost Benjamin Busch, who is the all-knowing fairy of the working group. Without Benni no IT-problem would have been solved, many mineral phases in a thin section would have remained mysterious, many a talk would have been dull, too long or too rapid. There could not have been a better postdoc and I could not be more thankful.

I was greatly supported by Yasar Manß, Ivy Becker and Lena Merz. All of whom helped setting up the new infrastructure of the working group, accompanied me for sampling and provided advice and constructive criticism whenever I got stuck. Thanks to them, drilling plug samples with outside temperatures of 0 to 35 °C and fighting off insects while fishing for plugs was always fun and adventurous. Other colleagues, who offered their insights and support were Alexander Monsees, Olajide Adamolekun and Rainer Babiel. I also profited greatly from project work conducted by the students Franziska Marquart, Benjamin Naumann and Benjamin Amar.

When coming to the KIT, I was given a warm welcome by Miriam Wiegand, Boris Reznik and Agnes Kontny. They introduced me to procedures at the KIT, facilities available within the geosciences and their specialty, rock magnetism, and thereby broadened my horizon.

Furthermore, I owe this thesis to Ulrike Brecht and Stephan Unrein. Ulrike, who kept her calm when facing the most obscure invoices and travel receipts and supported me also when Covid-19 diminished all travel arrangements for the TwinTraining project. And Stephan, who tackled with me new kinds of samples transitioning from hard rocks to bleached sandstones and even loose sediment so that I could end up with perfect thin sections. I am grateful to both of them for keeping the working group afloat and even more so for being such great people.

Most importantly I thank my family and friends, who offered me help in moments of struggle and advice in moments of doubt. Knowing that they cared as much as I did makes this thesis more precious to me.

Abstract

Reservoir quality in fluvial siliciclastic rocks is highly variable. Factors influencing permeability extend from micron-scale grain coatings, to mm-scale cemented and/or bleached horizons over pebble lags and cm-sized mud intraclasts to different styles of sorting and bedding that can persevere over meters or even kilometers. Outcrop analog studies allow the analysis of a specific depositional environment on the large scale and provide thin section samples to deduce the diagenetic processes including compaction, which act on the grain scale.

To unravel some of the reservoir quality controls in generally heterogeneous fluvial sandstones, this study defines the causes and consequences of the bleaching process as it occurs ubiquitously in terrestrial red beds. Such a process-oriented approach is also used to understand the lateral variability in a case study of a meander deposit and link the responsible phases in the mineral assemblage to continuously changing conditions during deposition in the river channel. Lithofacies types are explored with regard to their differing effects on compaction and cementation during burial and the consequential discrepancy between their porosity and permeability values.

Bleaching phenomena are investigated on the pore scale in an Upper Buntsandstein sandstone quarry. A 10 m thick red sandstone body is exposed in the outcrop with up to cm-sized, greyish white laminae in the sandstones. Analyses focus on bleached and unbleached zones in the same cm-sized samples.

Bleached zones exhibit a larger grain size (by 27 μm) as well as less compaction and a higher optical porosity (by 3 %). Stronger cementation by all observed authigenic phases of quartz, illite, K-feldspar and to a minor extent dolomite is registered during point counting. Calculated intergranular volumes and cementational porosity loss are positively linked with bleaching. Meanwhile unbleached zones have incorporated more ductile grains (e.g. micas, detrital clay, rock fragments like phyllites and plutonic rock fragments) and therefore undergo major porosity loss via compaction.

It can be shown that bleaching is related to primary features like grain size-selective lamination, which leads to higher permeability in coarse-grained laminae. In addition to that, it is relying on an early framework stabilizing cement phase, which maintains open pathways for uplift-related leaching of the detritus and few dolomite cements.

To assess the storage potential and permeability of a reservoir rock, lateral heterogeneity has to be taken into account. In the case study of a Lower Triassic outcrop the heterogeneity of the fluvial sandstones of the Plattensandstein Member (Röt Formation) near Röttbach in the southern part of the Germanic basin, Southern Germany, is investigated. The corresponding Buntsandstein Group is considered a suitable reservoir for thermal storage or exploration in Southern Germany. The >10 m thick lithic arkoses are well exposed over 35 m laterally in two perpendicular walls in the Röttbach Quarry and consist of two units separated by an erosive surface. They are covered by several meters of the Röt Claystone Member. Using sedimentological logging, spectral gamma ray-, porosity- and permeability measurements as well as petrographic analyses calculating compaction parameters, lateral reservoir quality differences are studied.

Sedimentary structures in the multi-storey channel suggest a point bar deposit, with the channel migrating in the direction corresponding to deteriorating reservoir qualities. This deterioration correlates with an increased detrital mica and authigenic illite content, the occurrence of rip-up clasts, and a lateral increase of gamma ray signatures by 19 API (avg. 5.0 % K, 2.7 ppm U, 11.9 ppm Th). The larger amounts of mica and illite are linked to decreasing flow velocities throughout the evolution of a meandering system. They enhance compaction and reduce available intergranular pore space. Illite grain coatings also inhibit diagenetic syntaxial quartz cement growth.

Permeability laterally deteriorates by up to three orders of magnitude (48 mD to 0.02 mD, avg. 9 mD) while measured porosities show minor variability (11 % to 19 %, avg. 16 %). The deterioration of reservoir quality is related to compactional (COPL 11 % to 28 %, avg. 18 %) and cementational porosity loss (CEPL 13 % to 29 %, avg. 22 %) and is mainly controlled by detrital mica and authigenic illite.

Reservoir quality is often related to depositional rock types and lithofacies. Within this study two lithofacies types from a Lower Triassic outcrop analog of the fluvial Vogesensandstein Formation at the eastern flank of the Upper Rhine Graben are analyzed. The corresponding Buntsandstein Group is a known reservoir lithology in the Upper Rhine Graben and Southern Germany, for geothermal energy, hydrocarbons, and potentially for storage. In the quarry in Lahr-Kuhbach 30 m high walls crop out featuring two lithofacies types of which 1.2 m of cross-bedded, medium-grained sandstones (St) and 0.8 m of fine-grained wavy- to ripple-bedded units (Swr) are the focus of this study to delineate reservoir quality controls. Both lithofacies types are analyzed with regard to sandstone structure, texture, porosity and permeability as well as mineralogy and its diagenetic evolution.

Lithofacies type St exhibits a slightly larger average grain size than Swr samples (by 0.05 mm) coinciding with lower spectral gamma ray measurements (by 19.9 API) and lower grain coating coverages (by 27 %). All of those parameters indicate slightly lower flow velocities of the river during Swr deposition leading to a higher input of detrital clays into the sediment. Smaller grain coating coverages and larger quartz cement volumes are found in the St samples and stabilize their grain framework as opposed to Swr samples. Grain contact types, intergranular volumes, and compactional porosity loss indicate a higher degree of chemical compaction within the Swr samples that correlates well with the higher grain coating coverages and smaller syntaxial overgrowth cement volumes. Compaction is the main control on reservoir quality leading to Swr samples exhibiting a lower average permeability by four orders of magnitude (0.09 mD vs. 236 mD).

Characterizing lithofacies types can thus offer a means to classify rock units according to differing styles of sedimentation and subsequent diagenetic overprint.

Kurzfassung

Die Reservoirqualität in fluviatilen, siliziklastischen Gesteinen kann stark variieren. Faktoren, welche das Fließverhalten beeinflussen, reichen von Kornbezügen im µm-Bereich, zementierten und/oder gebleichten Horizonten im mm-Bereich über Kieslagen und Tonintraklasten im cm-Bereich bis hin zu verschiedenen Arten der Sortierung und Schichtung, welche über Meter bis Kilometer vorliegen können. Analogstudien in Aufschlüssen ermöglichen die Analyse von Sedimentgesteinen spezifischer Ablagerungsbedingungen im großen Maßstab und liefern Dünnschliffproben, um die diagenetischen Prozesse einschließlich der Kompaktion abzuleiten, welche auf der Kornskala wirken.

Um einige der kontrollierenden Parameter für die Reservoirqualität in allgemein heterogenen fluviatilen Sandsteinen zu untersuchen, werden in dieser Studie die Ursachen und Folgen des Bleichungsprozesses herausgearbeitet, welcher häufig in terrestrischen Rotsedimenten auftritt. Ein solcher, prozessorientierter Ansatz wird gleichsam verwendet, um die laterale Variabilität in einer Fallstudie einer Mäanderablagerung zu verstehen und die verantwortlichen Phasen des Mineralgefüges mit den sich ständig ändernden Bedingungen während der Ablagerung im Flussbett in Verbindung zu bringen. Lithofaziestypen werden hinsichtlich ihrer unterschiedlichen Auswirkungen auf die Kompaktion und Zementation während der Versenkung und der daraus resultierenden Diskrepanz zwischen ihren Porositäts- und Permeabilitätswerten untersucht.

Bleichungsphänomene werden auf der Porenskala in einem Sandsteinbruch des Oberen Buntsandsteins untersucht. Im Steinbruch ist ein 10 m mächtiger Gesteinskörper bestehend aus rotem Sandstein aufgeschlossen, welcher bis zu cm-mächtige, grauweiße Laminen enthält. Die durchgeführten Analysen konzentrieren sich auf gebleichte und ungebleichte Zonen in cm-großen Proben.

Die gebleichten Zonen weisen eine gröbere Korngröße (um 27 µm) sowie weniger Kompaktion und eine höhere optische Porosität auf (um 3 %). Beim Point counting kann eine stärkere Zementation durch alle beobachteten authigenen Phasen, also Quarz, Illit, K-Feldspat und in geringerem Maße Dolomit, festgestellt werden. Die berechneten intergranularen Volumina und der Porositätsverlust durch Zementation korrelieren positiv mit der Bleichung. Zusätzlich haben ungebleichte Zonen vermehrt duktile Körner (z.B. Glimmer, detritischen Ton, Gesteinsfragmente wie Phyllite und plutonische Gesteinsfragmente) eingelagert und unterliegen daher einem größeren Porositätsverlust durch Kompaktion.

Es kann gezeigt werden, dass der Bleichungsprozess mit primären Merkmalen wie korngrößenselektiver Lamination zusammenhängt, welche zu höheren Permeabilitäten bei grobkörnigen Laminen führt. Darüber hinaus ist die Bleichung auf eine frühe, das Korngefüge stabilisierende, Zementphase angewiesen, welche Wegsamkeiten für die mit der Hebung verbundene Laugung des Detritus und der wenigen Dolomitzemente offenhält.

Um das Speicherpotenzial und die Permeabilität eines Speichergesteins zu beurteilen, muss dessen laterale Heterogenität berücksichtigt werden. In der Fallstudie eines Aufschlusses aus der Untertrias wird die Heterogenität der fluvialen Sandsteine der Plattensandstein-Subformation (Röt-Formation) bei Röttbach im südlichen Teil des Germanischen Beckens, Süddeutschland, untersucht. Die zugehörige Buntsandstein-Gruppe gilt als geeignetes Reservoir für die geothermische Speicherung oder Exploration in Süddeutschland. Die >10 m mächtigen lithischen Arkosen sind lateral über 35 m in zwei senkrechten Wänden im Steinbruch Röttbach aufgeschlossen und bestehen aus zwei Einheiten, die durch eine Erosionsfläche getrennt werden. Sie werden von mehreren Metern der Röttonsteine überlagert. Mit Hilfe sedimentologischer Profile, spektraler Gamma-Ray-, Porositäts- und Permeabilitätsmessungen sowie petrographischer Analysen zur Berechnung von Kompaktionsparametern werden laterale Unterschiede in der Reservoirqualität untersucht.

Die Sedimentstrukturen in den aufgeschlossenen, mehrstöckigen Flussrinnenablagerungen deuten auf eine Gleithangablagerung hin, wobei das Flussbett in die Richtung wandert, in der sich die Reservoirqualität verschlechtert. Diese Verschlechterung korreliert mit einem höheren Gehalt an detritischen Glimmern und authigenem Illit, dem Auftreten von Rip-up-Klasten und einer lateralen Zunahme der gemessenen Gammastrahlung um 19 API (durchschnittlich 5,0 Vol.-% K, 2,7 ppm U, 11,9 ppm Th). Die größeren Mengen an Glimmer und Illit stehen im Zusammenhang mit abnehmenden Strömungsgeschwindigkeiten während der Fortentwicklung des mäandrierenden Systems. Sie erhöhen die Kompaktion und verringern den verfügbaren intergranularen Porenraum. Illit-Kornbezüge hemmen wahrscheinlich zudem das Wachstum von diagenetischem Quarzzement.

Die Permeabilität verschlechtert sich lateral um bis zu drei Größenordnungen (48 mD bis 0,02 mD, Ø 9 mD), während die gemessenen Porositäten eine geringe Variabilität aufweisen (11 % bis 19 %, Ø 16 %). Die Verschlechterung der Reservoirqualität hängt mit dem Porositätsverlust durch Kompaktion (COPL 11 % bis 28 %, Ø 18 %) und durch Zementation (CEPL 13 % bis 29 %, Ø 22 %) zusammen und wird hauptsächlich durch detritische Glimmer und authigenen Illit kontrolliert.

Die Reservoirqualität hängt oft mit den durch die Ablagerung geprägten Gesteinstypen und der Lithofazies zusammen. Im Rahmen dieser Arbeit werden zwei Lithofaziestypen aus einem Aufschlussanalog der Untertrias aus der fluviatilen Vogesensandstein-Formation an der Ostflanke des Oberrheingrabens analysiert. Die entsprechende Buntsandstein-Gruppe ist ein bekanntes Reservoirgestein im Oberrheingraben und in Süddeutschland für geothermische Energie, Kohlenwasserstoffe und potenziell für Speicherung. Im Steinbruch in

Lahr-Kuhbach sind 30 m hohe Wände aufgeschlossen, welche zwei Lithofaziestypen beinhalten, von denen 1,2 m schräggeschichtete, mittelkörnige Sandsteine (St) und 0,8 m feinkörnige, wellenförmig horizontal- bis rippelgeschichtete Einheiten (Swr) im Mittelpunkt dieser Studie stehen, um Kontrollfaktoren für gemessene Reservoirqualitäten zu untersuchen. Beide Lithofaziestypen werden im Hinblick auf Sedimentstruktur, Textur, Porosität und Permeabilität sowie auf die Mineralogie und ihre diagenetische Entwicklung untersucht.

Lithofaziestyp St weist eine geringfügig größere durchschnittliche Korngröße auf als Swr-Proben (um 0,05 mm), was mit niedrigeren spektralen Gamma-Ray-Messungen (um 19,9 API) und niedrigerer Kutanabdeckung (um 27 %) einhergeht. All diese Parameter weisen auf leicht geringere Fließgeschwindigkeiten des Flusses während der Swr-Ablagerung hin, was zu einem höheren Eintrag von detritischen Tonen in das Sediment führte. In den St-Proben liegen geringere Kutanabdeckungen und höhere Quarzzementvolumina vor, letztere stabilisieren deren Korngefüge während das Gegenteil in Swr-Proben der Fall ist. Bevorzugt auftretende Korn-Korn-Kontakttypen, intergranulare Volumina und der Porositätsverlust durch Kompaktion weisen auf einen höheren Grad der chemischen Kompaktion innerhalb der Swr-Proben hin, welcher mit den höheren Kutanabdeckungen und den geringeren syntaxialen Quarzzementüberwüchsen korreliert. Die Kompaktion ist der wichtigste kontrollierende Faktor für die Reservoirqualität, was dazu führt, dass die Swr-Proben eine um vier Größenordnungen niedrigere durchschnittliche Permeabilität aufweisen (0,09 mD gegenüber 236 mD).

Lithofaziestypen zu charakterisieren bietet daher eine Möglichkeit Gesteinseinheiten gemäß ihren unterschiedlichen Ablagerungsbedingungen und ihrer nachfolgenden diagenetischen Überprägung zu klassifizieren.

Contents

Declaration of Originality

Chapter 2: As first author during her PhD studies, Christina Schmidt performed all sample acquisition, thin section analysis, point counting, compaction parameter calculations, data analysis, and interpretation. She wrote the entire article with inputs from Benjamin Busch and Christoph Hilgers. The discussions with all co-authors contributed to this manuscript.

Chapter 3: As first author, Christina Schmidt performed all sample acquisition, spectral gamma ray measurements, point counting, petrophysical measurements, data analysis, and interpretation. She wrote the entire manuscript with inputs from Benjamin Busch and Christoph Hilgers. The discussions with all co-authors contributed to this manuscript.

Chapter 4: As first author, Christina Schmidt performed all sample acquisition, spectral gamma ray measurements, point counting, petrophysical measurements, data analysis, and interpretation. She wrote the entire manuscript with inputs from Benjamin Busch and Christoph Hilgers. The discussions with all co-authors contributed to this manuscript.

1 Introduction

1.1 Motivation

As the 21st century advances, energy has become a pressing issue for the global society confronted with climate change and the rising energy demand of ever more developing countries. Renewables like solar and wind power have pushed the need to be able to store energy subsurface e.g. in the form of gas. At the same time carbon capture and storage has offered industries and countries a means to decrease their carbon footprint. In that context, reservoir rocks so far primarily essential for the oil and gas sector have after the emergence of geothermal energy gained once more in importance. In order to use reservoirs efficiently and on a large scale, reservoir characterization to determine their quality is indispensable. Properties affecting reservoir quality are the detritus and diagenetic phases but also processes like dissolution and compaction. During burial they alter the intergranular pore space available as reservoir for fluids and are thus relevant for production of water, gas or oil as well as for storage capacity.

Reservoir quality in a siliciclastic rock is affected by processes beginning with the erosion and transport of the detritus from the source area to the sedimentary basin, where it is deposited (Bjørlykke, 2014). Sandstone textures like grain size and roundness are first controls on the initial porosity and permeability because grain size and sorting affect the size and connectivity of the initial pore space (Füchtbauer, 1988c; Hsü, 1977; Wilson and McBride, 1988). Sandstone structures like cross-bedding, herringbone cross-stratification or wavy and ripple bedding resulting from specific transport energies and -mechanisms mostly impact permeability with special regard to directional permeabilities like vertical and horizontal permeability, k_v and k_h, and parallel or perpendicular to the bedding strike (Hartkamp-Bakker and Donselaar, 2009; Mast and Potter, 1963; Meyer and Krause, 2006; Pranter et al., 2007). Miall (1977b) introduced lithofacies types combining lithology and characteristic sedimentary structures to appropriately characterize fluvial facies in the field and identify the corresponding local depositional environment. Subsequently his lithofacies classification has been applied and extended by authors in the field (Eberth and Miall, 1991; Le Blanc Smith, 1980; Miall, 1996). The descriptive work of assigning a lithofacies type and determining its mode of deposition only sometimes has been related to measured porosities and permeabilities (Hornung and Aigner, 1999, 2004; Klingbeil et al., 1999; Wachutka and Aigner, 2001). The reasons for such eventual correlations between lithofacies type and porosity and/or permeability are still underexplored.

From detritus and depositional setting the causal chain can be continued to the subsequent processes acting on the rock during burial. Detrital clasts may affect diagenesis with respect to their resistance against dissolution providing ions for subsequent mineral precipitation and to compaction, both mechanical and chemical (Fig. 1.1) (Füchtbauer, 1988b; McBride, 1963). While feldspars and rock fragments are prone to early dissolution by meteoric waters, mineral grains like micas, which are not considered in standard detritus classifications (Folk, 1980; McBride, 1963), may also play an important role for mechanical and chemical compaction promoting the latter where in contact with quartz grains (Bjørkum et al., 1993; Sheldon et al., 2003). Lundegard (1992) introduced parameters calculated from point counting results that determine whether compaction or cementation is the prevailing process responsible for porosity loss (compaction porosity loss COPL, cementation porosity loss CEPL). Other authors adopted that technique but so far have not attempted to quantitatively correlate the parameters COPL and CEPL with specific rock properties (Ajdukiewicz and Lander, 2010; Ehrenberg, 1995; Lander and Walderhaug, 1999; Wang et al., 2012). Only qualitative observations in thin sections have been made linking e.g. mud clasts or bent micas with compaction (Fig. 1.1a) or pore-filling cements with limited mechanical compaction (Fig. 1.1b) (Salem et al., 2000). Using the scatter plot of COPL over CEPL could offer a means to evaluate correlations of compaction behavior with specific rock properties rather than assuming a link from petrographic analyses.

In addition to the detrital mineral assemblage, compaction of the intergranular volume is also known to depend on the presence of diagenetic phases as mentioned (Lundegard, 1992; Paxton et al., 2002; Salem et al., 2000). The first mineral phases forming after deposition are often grain coatings (Busch, 2020; Molenaar and Felder, 2018), they have been proven to facilitate chemical compaction (Monsees et al., 2020) and inhibit pore-filling cementation (Ajdukiewicz and Larese, 2012; Bloch et al., 2002; Busch, 2020; Busch et al., 2017; Lander and Walderhaug, 1999; Makowitz et al., 2006; Taylor et al., 2010). Grain coatings incorporating iron oxides cause the characteristic reddish brown rock color known from arid depositional settings (Walker, 1967b, 1975). Such rock color can be removed thereafter via a range of diagenetic processes including bleaching agents such as CO_2, organic matter or H_2S (Aehnelt et al., 2021; Beitler et al., 2003; Chan et al., 2000; Hilse et al., 2010). Bleaching is to be distinguished from leaching as the former is defined as a discoloration not necessarily coinciding with a reduction of the mineral volume, whereas the latter removes one or more mineral phases and thus mineral volume. If grain coatings are discontinuous, the formation of syntaxial overgrowth cements is not inhibited. Cementation is then dependent on the detrital substrate grains, e.g. quartz or feldspar, on the depositional environment, and a matching ion source. A lot is already known about the effect of the specific depositional setting onto diagenesis. Pore fluid chemistry for example is to a large extent controlled by the depositional setting, be it undersaturated fluid around a river channel dissolving grains or precipitating carbonate where the climate is semi-arid.

Or sulfate-, Fe-, and Ca-enriched sea water percolating through the sediment and precipitating carbonate and sulfate cements, while evaporitic cements form in (semi-)arid settings with pore fluids enriched in sulfate and respective cations like Ca^{2+} (Bjørlykke, 2014; Morad et al., 2010; Schreiber and El Tabakh, 2000). Quartz cements can be observed where enough silica is present in the pore fluid, e.g. from chemical compaction/pressure dissolution, to precipitate and an incomplete grain coating coverage allows for the growth of euhedral crystals (Lander et al., 2008; Walderhaug, 1996). In the past they have often been described as main cause for reservoir quality deterioration by pore-filling cementation and best prevented by grain coatings like chlorite (Ajdukiewicz and Lander, 2010; Ajdukiewicz and Larese, 2012; Ajdukiewicz et al., 2010; Bloch et al., 2002; Taylor et al., 2010; Walderhaug, 1996). But some authors have registered a stabilization of the grain framework by quartz cement, which in turn preserves intergranular pore space (Fig. 1.1b) (Makowitz et al., 2006; Molenaar, 1986; Walderhaug, 2000). It seems thus that the impact of grain coating coverage and quartz cement volumes onto reservoir quality has to be evaluated independently with regard to compaction parameters. A procedure that so far is rarely undertaken and should be included into reservoir characterization.

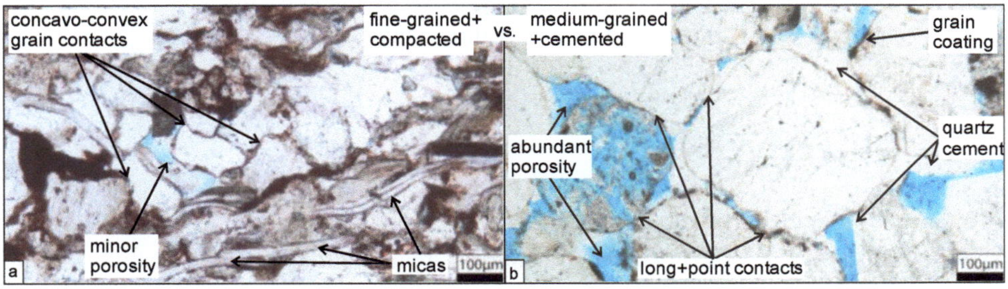

Figure 1.1: a) Example of a more fine-grained facies with abundant iron-rich coatings and -pseudomatrix. Porosity is low and the grain framework has been strongly compacted where grain contacts are concavo-convex. Micas have been subjected to mechanical deformation. b) Example of a more coarse-grained facies with rare grain coatings preventing pore-filling cementation. Instead quartz cement partially fills the pore space but primarily stabilizes the grain framework and available porosity. Compaction is less pronounced as can be seen from prevailing point and long grain contacts.

1.2 Aims

This study aims at a better understanding of the interplay between primary sandstone composition beyond quartz, feldspar, and rock fragments, texture and structure with subsequent compaction, and diagenesis throughout burial. Individual rock properties like grain size, mica content, grain coating coverage, authigenic quartz cement volume are analyzed with special regard to their influence on porosity and permeability but also on compaction behavior. The construction of paragenetic sequences allows for those properties to be put into a relative chronological order of reservoir quality controls, especially in the case of bleaching in the Lower Triassic red beds. The role of compaction for reservoir quality is investigated in detail in the targeted fluvial outcrop analogs, which are known reservoir lithologies in Southern and Central Germany. Accordingly, this work's focus is on the following aspects:

- Showing that bleaching is not causing improved permeabilities in the subsurface but rather benefits from them during the bleaching process. Still bleaching can be a valid indicator for zones of higher reservoir quality.
- Coupling lateral variability of reservoir quality within a point bar deposit with progressing channel migration and a concomitant reduction in flow velocity of the river stream leading to an enhanced incorporation of mica and fines into the deposit.
- Underlining the significance of often neglected mica and rip-up clasts for compaction behavior and in turn reservoir quality.
- Pointing out the importance of individually assessing grain coating coverage and quartz cement volume in every outcrop or subsurface target to evaluate whether porosity and permeability reduction are primarily the consequence of compaction promoted by clay minerals or of pore-filling cementation.

1.3 Objectives

In order to elaborate correlations existing in the targeted Buntsandstein outcrops, petrophysical properties, i.e. porosity and permeability, are measured on all plug samples. Mineral assemblages are studied petrographically in order to determine sample texture, detrital composition and mineral alterations resulting from diagenetic processes.

Regarding the Upper Buntsandstein outcrop in the active quarry in Röttbach, 43 plugs are sampled, originating both from the quarry wall (N = 36) and from mined blocks (N = 7). An equal number of thin sections is available for petrographic analysis (e.g. Fig. 1.2). Bedding allowed for the reconstruction of sample orientation in the mined blocks. Eleven samples are (partially) bleached and a comparative workflow can be applied to delineate essential differences in mineralogy and optical porosity within a single sample. Of the unbleached samples, 31 are petrographically assessed and used as calibration for the unbleached zones, while the two fully bleached samples offer a means to calibrate the bleached zones of such partially bleached samples. The compaction state is analyzed based on grain contact types after Taylor (1950) and the intergranular volume calculated after Paxton et al. (2002) in order to aid in determining the paragenetic sequence, timing of bleaching and its impact on primary porosity.

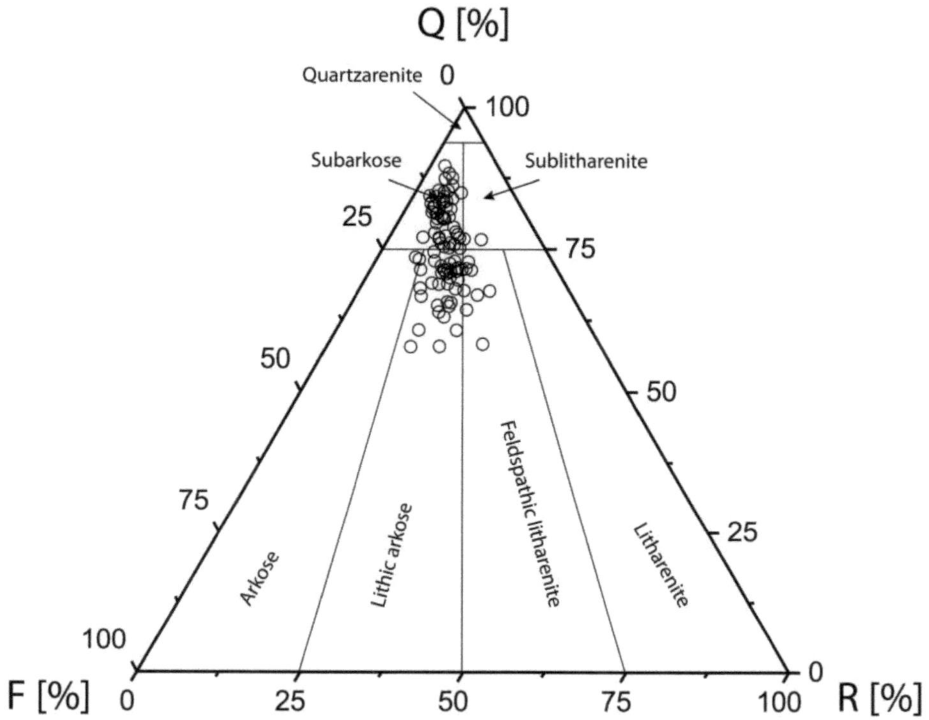

Figure 1.2: Overview of the detrital compositions as they were determined for all samples from thin section-analyses.

In a second study, the lateral variability of the fluvial outcrop Röttbach is examined. Therefore, the samples from the quarry wall (N = 36), which are used in the first study to investigate bleaching phenomena, are regarded with special focus on their position along the three profiles that were sampled in two vertically stacked sedimentary units. The profiles extend over 35 m and cover walls of differing orientations, with two samples per profile from the SE-quarry wall and as well as two samples per profile from the SW-wall. Remaining samples were derived from the NW-wall. Two profiles originate from the basal sandstone unit and one from the top sandstone unit. All samples were drilled horizontally into the quarry wall. The two sandstone units are separated by a subhorizontal erosive surface. Such surfaces and sedimentary bedding are classified after Miall (1988a, 1988b) in order to understand conditions during deposition and deduce river channel migration. In this study no distinction of bleached and unbleached zones is undertaken, samples undergo petrographic analyses as bulk samples but neglecting two samples containing large intraclasts, which would bias overall results from the optical assessment (N = 34, e.g. Fig. 1.2). Via porosity and permeability measurements (N = 36, e.g. Fig. 1.3) and petrographic analyses including compaction after Taylor (1950) and Paxton et al. (2002) depositional factors are highlighted that influence the diagenetic evolution and thereby reservoir quality.

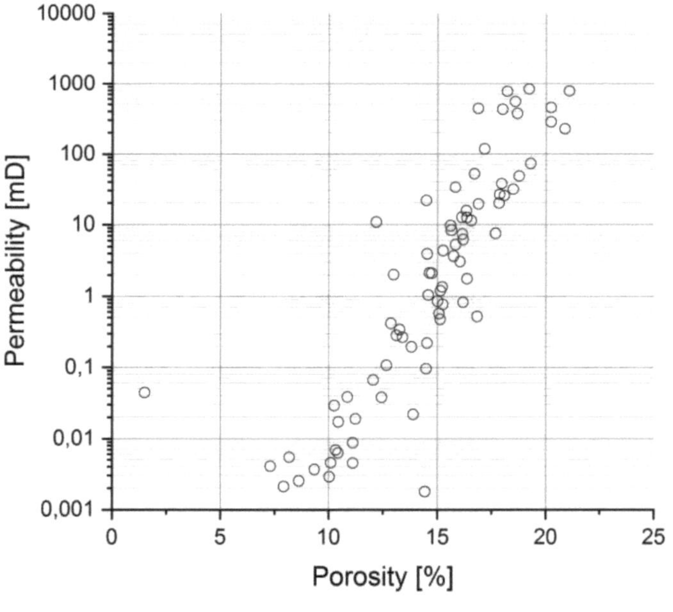

Figure 1.3: Overview of all porosity and permeability values that were determined in the rock samples from the two outcrop localities.

The third study focusses on the impact of lithofacies on reservoir quality in the active quarry Lahr-Kuhbach. The featured Vogesensandstein Formation (Middle Buntsandstein) of the marginal facies of SW-Germany is investigated with regard to its mineral composition, diagenetic alteration and petrophysical properties. The outcrop is characterized by two lithofacies types after Miall (1977, 1996) that form a recurring succession of trough cross-bedded (St) and wavy- to ripple-bedded sandstones (Swr). Samples of three orientations can be taken due to the perpendicular quarry walls striking N-S and W-E. Thus, overall 23 St samples (15 horizontal, 8 vertical plugs) and 19 Swr (12 horizontal, 7 vertical plugs) are taken. All 42 samples are analyzed petrographically to determine microscopic differences in e.g. mineralogical composition (e.g. Fig. 1.2), grain size, grain coating coverage or diagenetic mineral content characterizing the macroscopically defined lithofacies types. Petrophysical measurements of porosity and permeability (e.g. Fig. 1.3) and the compaction after Taylor (1950) and Paxton et al. (2002) are correlated with mineralogical characteristics. This way it is evaluated whether the descriptive lithofacies nomenclature allows for an approximation of reservoir quality.

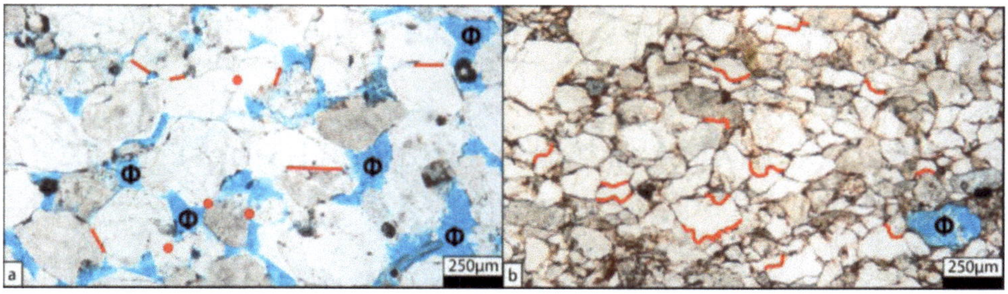

Figure 1.4: Comparative overview thin sections illustrating how different compaction states correlate with available open pore space. a) Minor degrees of compaction associated with primarily point and long contacts (red points and lines) and coinciding with abundant open, primary porosity (Φ) despite significant, syntaxial quartz cement overgrowths. b) Intense compaction forming concavo-convex to sutured grain contacts (red) allowing only for minor remaining porosity (Φ), in this case from grain dissolution, and barely authigenic cement volumes.

1.4 Overview of the thesis

1.4.1 Compaction and cementation control on bleaching in red beds (Chapter 2)

Bleaching is a well-known phenomenon in red bed siliciclastics. It can have various causes all leading to a reducing environment of low pH and high salinity, e.g. the presence of organic matter (Chan et al., 2000; Parry et al., 2004; Schöner and Gaupp, 2005; Weibel, 1998). Iron reduction and mobilization mostly affects iron oxides like hematite, which is mainly responsible for the red color of the deposits (Chan et al., 2000; Parry et al., 2004). Bleached zones have been found to be related to very specific depositional environments (Pudlo et al., 2012) as well as occurring regardless of the depositional setting (Beyer, 2015) or bound to the presence of bitumen (Schöner and Gaupp, 2005). It has been linked to grain size differences and an early pore-filling cement phase (Beyer, 2015; Kunkel et al., 2018). What remained unclear so far is whether porosity and permeability correlate with bleaching (Eichhubl et al., 2004; Pudlo et al., 2012; Reinecker et al., 2015) or are more or less decoupled from it (Kunkel et al., 2018; Schöner and Gaupp, 2005).

In chapter 2 petrography is used to identify characteristics of bleached zones compared to unbleached zones within single plug samples. We show how grain size differences inherent to lamination subsequently control the precipitation of early pore-filling phases preventing compaction and enabling a preferential flow of a reducing agent responsible for the bleaching. It can be stated that bleaching therefore is not a factor greatly enhancing the intergranular pore space and thus fluid pathways. However, it points at zones of higher reservoir quality resulting from the deposition of a more coarse-grained lithology. Samples were derived in an active quarry of the Lower Triassic Plattensandstein Member in Röttbach, Germany, belonging to the Bamberger Natursteinwerke Hermann Graser GmbH.

1.4.2 Lateral variations in an outcrop analog of Triassic fluvial red beds (Chapter 3)

Vertical and lateral variability in fluvial sandstones is a risk in reservoir quality prediction. It is by no means recognizable neither in a 1D well log or core nor in 2D inter-well correlations as for example architectural elements like lateral or downstream accretion deposits are too small to be detected in between two wells (Becker et al., 2017; Bridge, 2006; Miall, 1977a, 1988a; Miall and Turner-Peterson, 1989; Pranter et al., 2007; Taylor and Ritts, 2004). While sedimentological variability in an outcrop can often be recognized, associated rock properties causing the variability in petrophysical measurements on the grain scale are difficult to identify (Becker et al., 2017; Doyle and Sweet, 1995; Hartkamp-Bakker and Donselaar, 2009; McKinley et al., 2011).

In chapter 3 an outcrop analog study has been conducted following three profiles of different stratigraphic heights in a Lower Triassic sandstone quarry. Samples were all derived from two sandstone bodies representing deposition by lateral accretion. While measured porosities do not differ remarkably, permeabilities uniformly decrease along all three profiles in direction of the apparent dip in the respective quarry walls. Next to sedimentological observations, detrital components, and textural parameters like grain size as well as the prevailing authigenic phases like quartz, illite, and feldspar were analyzed. That way we can show that decreasing river flow velocities during continuous channel migration lead to changing mineral contents in the rock, which are registered by the spectral gamma ray log. They are influential enough to enhance chemical compaction and decrease permeabilities. Samples were derived in an active quarry in Röttbach, Germany, belonging to the Bamberger Natursteinwerke Hermann Graser GmbH.

1.4.3 Lithofacies control on reservoir quality in Triassic fluvial red beds (Chapter 4)

Lithofacies types are a classification scheme that is widely used to describe deposits, especially fluvial ones, in the field based on lithology and sedimentary structure (Eberth and Miall, 1991; Le Blanc Smith, 1980; Miall, 1977b, 1996). They are used in a descriptive manner and are interpreted with regard to the depositional setting in which they formed (Miall, 1977b, 1996). Authors who correlated them with petrophysical properties, so far did not have a closer look at the exact rock properties of the specific type that were responsible for observed trends in porosity and permeability (Hornung and Aigner, 1999, 2004; Wachutka and Aigner, 2001).

In chapter 4 two sandy lithofacies types from a Buntsandstein outcrop in the Middle Black Forest of Southern Germany are analyzed. The two types exhibit distinct differences in porosity and permeability that appear to be decoupled from minor grain size differences. Spectral gamma ray readings point towards notably different clay mineral contents. We can confirm that where clay mineral contents appear lower, grain coating coverage is less continuous and quartz cement volumes are higher, which coincides with higher point counted porosities. COPL-CEPL plots after Lundegard (1992) underline that both rock properties are of major influence to compaction, the intergranular volume, and at the same time controlling reservoir quality. We can thus state that minor changes in depositional setting causing a change in lithofacies type can have a significant impact on petrophysical properties within a rock that is otherwise considered as one coherent reservoir unit.

1.5 Parts of this thesis which have been published

Articles

Schmidt, C., Busch, B., Hilgers, C., Compaction and cementation control on bleaching in Triassic fluvial red beds, S-Germany, Journal of Applied and Regional Geology, Vol. 171(4).

Schmidt, C., Busch, B., Hilgers, C., Lateral variations of detrital, authigenic and petrophysical properties in an outcrop analog of the fluvial Plattensandstein, Lower Triassic, S-Germany, Journal of Applied and Regional Geology, Vol. 172(1).

Submitted manuscripts

Schmidt, C., Busch, B., Hilgers, C., Lithofacies control on compaction and reservoir quality in Lower Triassic fluvial red beds, S-Germany.

Conference contributions

Schmidt, C., Busch, B., Hilgers, C., How lithofacies types in a reservoir analog facilitate reservoir quality assessment. Geoscience & Engineering in Energy Transition Conference, Strasbourg (France), 2020.

Schmidt, C., Busch, B., Hilgers, C., Bleaching phenomena and their link to compaction in fluvial reservoir formations – Lower Triassic, Central Germany. SPE Student Technical Conference, Karlsruhe (Germany), 2020.

Schmidt, C., Busch, B., Hilgers, C., Lateral heterogeneity of reservoir properties in a homogenous Lower Triassic sandstone, Southern Germany. DGMK/ÖGEW Frühjahrstagung, Celle (Germany), 2019.

Schmidt, C., Busch, B., Hilgers, C., Coupling diagenesis and petrophysics in red bed siliclastics to infer fluid flow anisotropies - Analog studies from SW Germany, EAGE Conference and Exhibition, Kopenhagen (Denmark), 2018.

2 Compaction and cementation control on bleaching in Triassic fluvial red beds, S-Germany

2.1 Abstract

This study focusses on bleaching phenomena on the pore scale in an Olenekian (Upper Buntsandstein) sandstone quarry. The study area exposes a 10 m thick red sandstone body with up to cm-sized, greyish white laminae in sandstones. Analyses focus on bleached and unbleached zones in the same cm-sized samples.

Bleached zones show a larger grain size (by 27 µm), less compaction, and a higher porosity (by 3 %). They also exhibit stronger cementation by all observed authigenic phases of quartz, illite, K-feldspar, and to a minor extent dolomite. Calculated intergranular volumes and cementational porosity loss also correlate positively with bleaching. Meanwhile unbleached zones contain more ductile grains (e.g. micas, detrital clay, rock fragments like phyllites and plutonic rock fragments) and are affected by major porosity loss via compaction.

Bleaching is related to primary features like grain size-selective lamination and associated higher permeability in coarse-grained laminae. It is also reliant onto an early framework stabilizing cement phase, which keeps pathways open for uplift-related leaching of the detritus and few dolomite cements.

2.2 Introduction

Various authors have described bleaching (discoloring) in continental red beds, reflected by the disappearance of the sediment's original red color. Bleached specimens have been described from Lower and Middle Permian (Gaupp et al., 1993; Hilse et al., 2010; Pudlo et al., 2012; Schöner and Gaupp, 2005), Lower Triassic rocks in Central Europe (Beyer et al., 2014; Beyer et al., 2011; Hilse et al., 2010; Kasch et al., 2010; Kunkel et al., 2018; Wendler et al., 2012), Jurassic rocks in Utah and Nevada, USA (Beitler et al., 2003; Chan et al., 2000; Eichhubl et al., 2004; Parry et al., 2004), and elsewhere. Such phenomena may be pervasive or lamina-bound bleaching, or may have a speckled appearance or halos around clay clasts (Beitler et al., 2003; Beyer, 2015; Chan et al., 2000; Hilse et al., 2010; Pudlo et al., 2012; Schöner and Gaupp, 2005). Tectonic structures like joints, faults, and deformation bands may also be associated with bleached lithologies (Busch et al., 2017; Eichhubl et al., 2004; Kasch et al., 2010; Parry et al., 2004; Wendler et al., 2012). The sediment's red color originates from red pigments forming pigmented dust rims or continuous coatings around detrital grains. The red color forms via the alteration of iron-bearing grains and infiltrated iron-bearing clay under oxidizing Eh-conditions and neutral to alkaline pH in situ and after the settling of the sediment, respectively (Amelung et al., 2018; Walker, 1967a, b; Walker et al., 1978; Weibel, 1998).

The process of bleaching may be a process initiated by the migration of fluids causing a reducing environment at low pH and elevated salinities (Chan et al., 2000; Parry et al., 2004; Schöner and Gaupp, 2005; Weibel, 1998). Potential reducing and thus bleaching agents are hydrocarbons, methane, organic acids, and hydrogen sulfide, which lower the redox potential of the fluid that subsequently reduces the hematite of the formation. Depending on the reactants, CO2, water, and sometimes pyrite are produced (Chan et al., 2000; Eichhubl et al., 2004; Garden et al., 2001; Gaupp et al., 1993; Gaupp and Schöner, 2008; Haszeldine et al., 2005; Parry et al., 2004; Schöner and Gaupp, 2005; Shebl and Surdam, 1996; Weibel, 1998). The reduction of hematite leads to an increase in mobile Fe2+ in solution that is then available for local removal. At higher Eh of the pore fluid, a proximal, subsequent oxidation to Fe3+ and reprecipitation as iron (and manganese) oxide concretions is possible (Chan et al., 2000; Parry et al., 2004). A reduction of hematite can also be triggered by microbial activity with the organisms' enzymes reducing the metal species (Dong et al., 2014; Lovley et al., 1993). Alternatively, bacteria can reduce sulfate consuming organic matter and forming e.g. H2S, which in turn reduces hematite (Kirkland et al., 1995; Machel, 2001; Schumacher, 1996). Bacterial sulfate reduction is commonly replaced by thermochemical sulfate reduction at higher subsurface temperatures but also consumes organic matter (Machel, 2001). Similarly, pedogenesis may lead to color mottling where e.g. stagnant groundwater locally constituted a reducing microenvironment (Amelung et al., 2018; Buurman, 1980; Molenaar, 1984). Wendler et al. (2012) also associate CO2-rich fluids with bleaching and deduce a volcanic or meteoric origin. CO2 influxes may cause a pH-decrease contributing to iron mobilization (Schöner and Gaupp, 2005).

Pudlo et al. (2012) describe bleaching on a regional scale to be restricted to specific depositional settings like fluvial channel fills and eolian dunes rather than sandflat or even lacustrine deposits, while Beyer (2015) observes bleaching principally in e.g. sandflat deposits. Beyer (2015) relates bleaching to factors like grain size and cement content promoting fluid-rock-interactions. For the Lower Triassic in Central Germany bleaching was associated with medium-grained sandstones with early blocky cements, mostly carbonates and sulfates, which occur mostly in cross-bedded sandstones and can be interpreted as sandsheet or channel fill facies (Kunkel et al., 2018). Schöner and Gaupp (2005) deduce from bitumen staining that bleaching is not facies-bound but merely correlates with bitumen impregnation or as proposed by Beyer (2015) with acidic fluids preceding hydrocarbon emplacement. However, no consistent model can be established. While some studies show that bleaching, petrophysical properties, and lithofacies are related (Beyer, 2015; Eichhubl et al., 2004; Pudlo et al., 2012; Reinecker et al., 2015), other data neither suggest a correlatable impact of those properties on bleaching nor do they deny it (Beyer, 2015; Kunkel et al., 2018; Schöner and Gaupp, 2005).

In this study, we present data on a Lower Triassic red bed deposit with different amounts of decoloring, which is attributed to bleaching. We discuss lamina-bound and speckled decoloring with respect to sandstone petrology attempting at a better understanding of textural characteristics influencing the bleaching mechanism and evaluate its effect on reservoir properties.

2.3 Geological setting

The study area Röttbach is located 5 km N of the town Wertheim, Southern Germany, and is part of the German Triassic (Hagdorn et al., 2009; Meschede and Warr, 2019; Okrusch et al., 2011). According to the regional stratigraphy, the Lower Triassic deposits in the Röttbach quarry belong to the Plattensandstein (Röt Formation) of the late Olenekian (Upper Buntsandstein) (Menning, 2018; Okrusch et al., 2011; Röhling et al., 2018) (Fig. 2.1). They were deposited as marginal facies at the southern border of the North German Basin (Freudenberger et al., 2013; Hagdorn et al., 2009; Okrusch et al., 2011; Ziegler, 1990). Source areas are primarily the Vindelician High of Southern Germany and Switzerland to the SE and to a minor extent the Ardennes and Gallic Massif of France and Belgium to the W (Freudenberger et al., 2013; Hagdorn et al., 2009; Meschede and Warr, 2019; Röhling and Lepper, 2013; Ziegler, 1982). Correlations of Buntsandstein sequences across the marginal facies of Southern Germany are detailed in Fig. 5 by Röhling et al. (2018).

The Lower Triassic sandstones evolved from braided geometries in the Induan and early Olenekian (Early and Middle Buntsandstein), towards occasionally ephemeral, meandering streams during the late Olenekian to Anisian (Upper Buntsandstein) (Freudenberger et al., 2013; Okrusch et al., 2011). Those meanders are described to have a relatively constant position, low incision depth but may meander over a large width (Freudenberger et al., 2013; Okrusch et al., 2011). The late Olenekian Lower Röt claystones have been deposited in a sabkha to playa environment (Okrusch et al., 2011). They consist of silt- and claystones and can contain salt or sulfate nodules (Freudenberger et al., 2013; Mader and Teyssen, 1985; Okrusch et al., 2011).

The Plattensandstein, the focus of this study, has a thickness of up to 34 m in the study area (Freudenberger et al., 2013). Triassic Olenekian rocks are overlain by Anisian sediments, and locally crosscut by Upper Cretaceous to Tertiary basaltic and phonolitic volcanic dykes in the study area (Okrusch et al., 2011). The nearest reported exposure is a Tertiary explosive tuff chimney of 200 x 150 m diameter situated 31 km WNW of the study area (Okrusch et al., 2011).

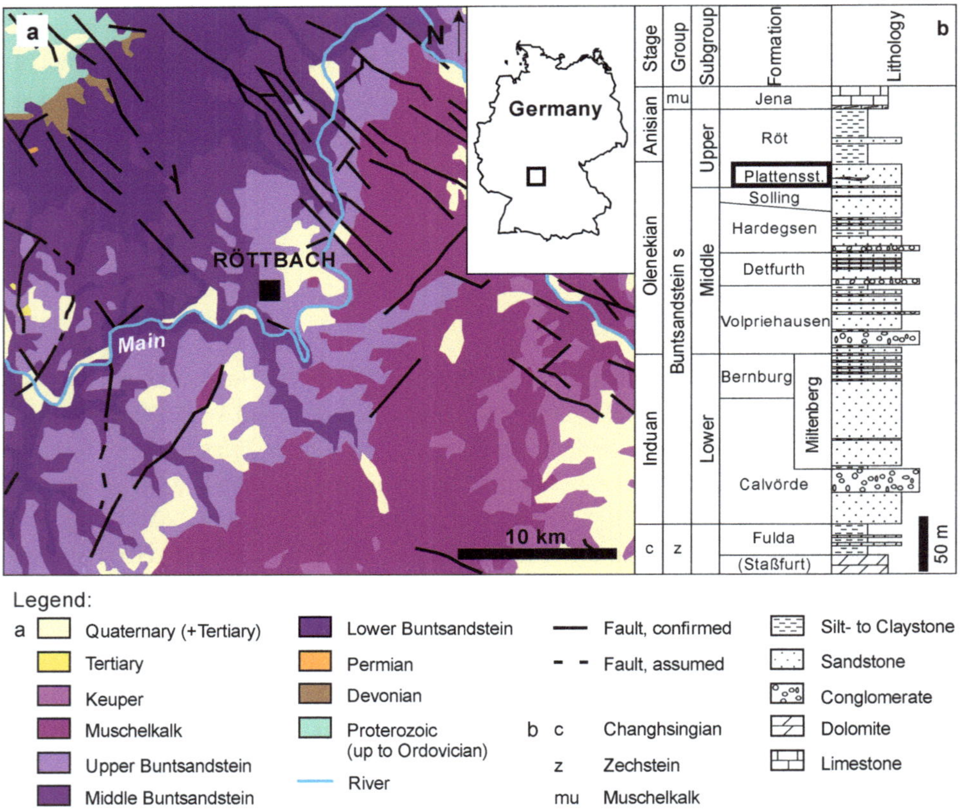

Figure 2.1: a) Location of the study area on the geological map of the SE-margin of the Spessart (after Okrusch et al., 2011). b) Lithological column displaying the stratigraphic position of the Plattensandstein, late Olenekian, (black box) in the footwall of the Lower Röt Claystone Member, Röt Formation (Upper Buntsandstein Subgroup) after Okrusch et al. (2011) modified from Geyer (2002).

2.4 Materials and methods

2.4.1 Sampling

The sandstone samples were drilled from the quarry wall (N = 36) as well as from mined m-sized blocks with distinctive bedding (N = 7). The 43 one-inch plugs were cut to lengths between 2.5 to 4.0 cm and oven-dried at 40°C for 7 days. Thin sections were prepared from resulting plug caps. Of those 43 samples, only eleven exhibit bleaching phenomena, two of them are fully bleached (RB-06, RB-07, Fig. 2.2, Tab. A.1). The nine partially bleached samples, containing lamina-bound and speckled bleaching, were used to compare characteristics of the unbleached and bleached material within one sample (RB-04, RB-05, RB-17, RB-24, RB-26, RB-28, RB-35, RB-38, RB-42). All other samples were completely unbleached and are used as a reference for unbleached zones in samples with bleaching.

Figure 2.2: Examples for different grades of bleaching in plugs. Partial bleaching occurred primarily lamina-bound. Displayed is the range from unbleached brownish red (a, sample RB-30) to partially bleached, yel-lowish to greyish white samples (b, sample RB-28), (c, sample RB-38), (d, sample RB-42), and fully bleached yellowish white samples (e, sample RB-06).

2.4.2 Petrography and porosity

Plug caps were impregnated with blue-dyed epoxy and thin sections prepared from them with a thickness of 30 µm, nine samples were stained with a combination of Alizarin Red S and potassium ferricyanide to identify carbonate cements.

Grain size was measured separately for both unbleached and bleached zones. Grain size analysis is done on a Leica DMLP microscope determining the length of the long axes of 100 detrital grains per thin section on a grid adjusted to the largest grain size to gain area-weighted results of the apparent grain size in 2D (Becker et al., 2017; Busch et al., 2019; Busch et al., 2020; Busch et al., 2017; Walderhaug, 2000; Walderhaug et al., 2012; Wüstefeld et al., 2017). Dust rims of hematite pigments trace the original detrital grains and are used to distinguish them from authigenic cement and assess the detrital grain size (Busch et al., 2017). Subsequently, the grain size class was determined according to the Udden-Wentworth scale (Udden, 1914; Wentworth, 1922) and the degree of sorting was calculated using the formula by Folk and Ward (1957). The bias of the measurement introduced when measuring on 2D sections is approximated in two samples (RB-10, RB-28) when measuring the grain's long axis with or without the dust rim on 20 grains. Measuring the dust rim as part of the grain resulted in an average grain size larger by 6 to 8 µm. Grain morphology for both bleached and unbleached zones was assessed using the comparator after Füchtbauer (1988c).

Point counting was performed on a Leitz Aristomet transmitted light microscope, fitted with a semi-automatic Pelcon Point counting stage and the grid adjusted to the maximum grain size. The specific modal composition was analyzed separately for both unbleached and bleached zones (300 steps for each zone) in the nine samples exhibiting partial bleaching. Uniformly colored thin sections and both unbleached and bleached zones in partly discolored thin sections were point counted with 300 steps.

Each lithoclast or rip-up clast was counted once and their internal components were not considered during point counting. Two samples contained rip-up clasts of several mm in diameter (RB-17, RB-19). Those were neglected during point counting of the bulk thin section but RB-17 was included in the selective point counts of the unbleached and bleached zone. Iron oxide phases (pore-lining and filling) are point counted as a detrital phase, since their ductile deformation due to compaction makes a clear distinction of authigenic and detrital iron oxide phases difficult. Dust rims were used for the distinction between authigenic and detrital phases of quartz (Busch et al., 2017). Comparison of cement volumes deduced from transparent thin section and cathodoluminescence analysis shows that cement volumes can accurately be assessed by transmitted light microscopy (Makowitz and Sibley, 2001; Walderhaug, 2000). Results for the detrital composition are plotted in a QFR-plot according to Folk (1980) with normalized amounts of detrital quartz (Q), feldspar (F), and rock fragments (R).

The porosity was determined separately for both unbleached and bleached zones in the partially bleached samples by point counting. Microporosity was calculated from point counts based on mean fractions of microporosities in kaolinite, chlorite, and illite after Hurst and Nadeau (1995) and the average of those values for undifferentiated clay minerals. The sum of the calculated microporosity and point counted primary and dissolution pore space constitutes the total point counted porosity.

The type of grain-grain contacts and the reduction of initial porosity can quantify compaction. Grain contact description follows the classification of Taylor (1950) as tangential or point (P-), long (L-), concavo-convex (C-), and sutured (S-) contacts and is applied to bleached and unbleached zones, reflecting increasing compaction (Taylor, 1950). Accordingly, grains in loosely compacted sediments without contacts are named floating grains (F-contacts). To quantitatively derive the degree of compaction those five categories have been point counted among 100 grains in both unbleached and bleached zones. For each grain the contact was counted, which exhibited the strongest indication for compaction. Additionally, the contact strength was calculated for grain-to-grain contacts as a measure of compaction with higher values representing a higher degree of com-paction (Füchtbauer, 1967). Intergranular volume (IGV) is calculated from point counting data according to Paxton et al. (2002). The IGV is the sum of the intergranular pore space, intergranular cement volumes, and depositional matrix, it is indicative of the degree of compaction (Paxton et al., 2002). Subsequently, further parameters can be determined based on the intergranular volume or minus-cement porosity as defined by Lundegard (1992), which is optical porosity plus intergranular cement volume. This includes porosity loss due to compaction (COPL) and porosity loss due to cementation (CEPL) (Lundegard, 1992). The initial porosity is assumed to be 45 % according to Wilson and McBride (1988) for unconsolidated sand in a fluvial environment. Porosity loss by compaction is plotted against cementational porosity loss (Lundegard, 1992).

2.5 Results

2.5.1 Bleached versus unaltered material

Macroscopic description, texture, and structure

Bleaching is primarily lamina-bound but not strictly limited to a single lamina (see Fig. 2.2b, c). Such bleached laminae are up to 1 cm thick and extend over ≥ 1 m. Of the eleven bleached samples, two are completely bleached (e.g. Fig. 2.2e), two exhibit bleached laminae of various thickness (up to 8 mm thick, Fig. 2.2b, approx. 1 mm thick in Fig. 2.2c). The remaining seven display speckled but still bedding-parallel bleaching up to 3 mm in diameter (Fig. 2.2d). The weathered outcrop walls barely show bleaching, bleaching was visible in mined blocks and after sampling. No indication of a specific propagation direction of the decoloration apart from along laminae is given. Pre-cipitation of iron phases either by abiotic components (Liesegang) or by biotic components (iron-oxide cement bands) (Burgess et al., 2016) was not encountered.

For the eleven samples exhibiting bleaching, the average grain size of bleached zones is 173 µm (max. 185 µm, min. 151 µm), while the respective unbleached zones in the same samples have an average grain size of 146 µm (max. 163 µm, min. 130 µm, Fig. 2.3a). In each of those samples a discrepancy of 16 to 40 µm between the average grain size of the bleached and unbleached zone is found (Fig. 2.3b). The fully bleached samples both have average grain sizes of 176 and 182 µm (sample RB-06 and RB-07, Tab. A.1).

In samples with bleached laminae, all unbleached zones are moderately sorted. For the bleached zones and fully bleached samples, four of eleven samples are well sorted. The grains' roundness does not show any distinctive changes between unbleached and bleached zones. Both contain subangular to subrounded grains (Fig. 2.6a, b, Tab. A.1).

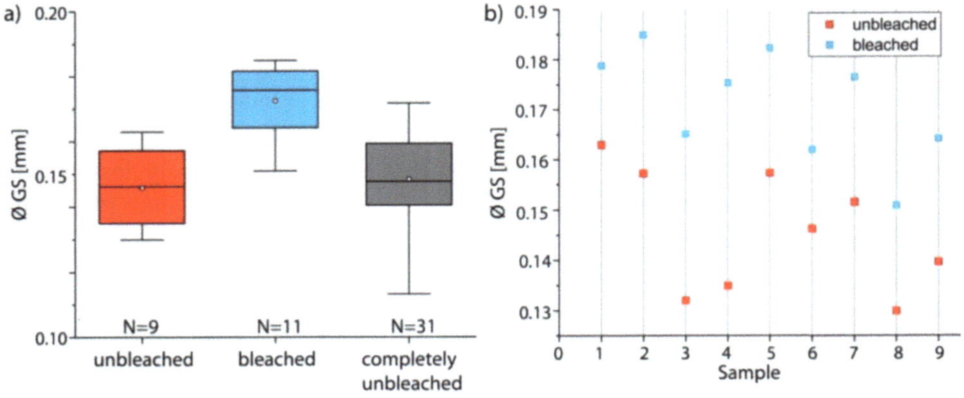

Figure 2.3: a) Box-whisker plot of average grain sizes (Ø GS) plotted for the three types of material: red unbleached zones, white bleached zones and completely red thin section samples (grey). The mean average grain size of bleached zones (avg. 173 µm) exceeds the unbleached zones (avg. 146 µm) by 27 µm. Completely unbleached samples cover the grain size range of all unbleached zones and reach a maximum grain size similar to the mean average grain size of the bleached zones. b) The scatter plot for the nine partially bleached, individual samples shows that bleaching in each sample coincides with a constantly larger average grain size.

Petrography

Quartz grains constitute equally high fractions of the overall mineral assemblage in both unbleached (avg. 35 %, max. 43 %, min. 27 %) and bleached zones (avg. 38 %, max. 46 %, min. 28 %) (Fig. 2.4). Detrital K-feldspar grains account for on average 8 % (max. 11 %, min. 4 %) in unbleached zones and for 11 % (max. 16 %, min. 7 %) in bleached zones. Rock fragments are present with 7 % (max. 11 %, min. 4 %) in unbleached zones. In bleached zones they make up 7 % (max. 11 %, min. 4 %) of the rock.

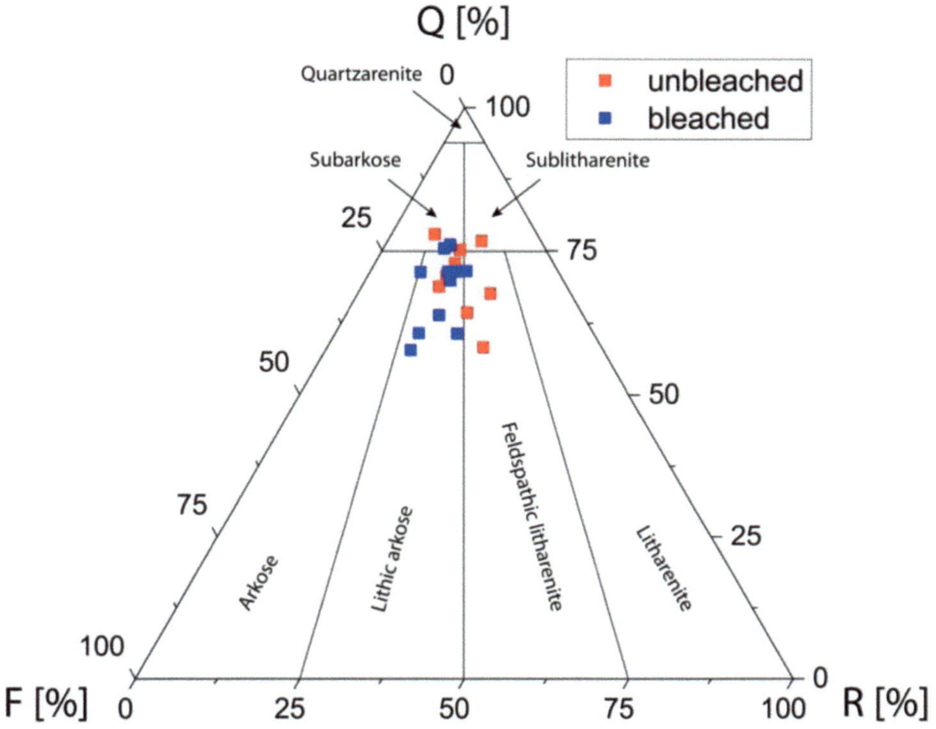

Figure 2.4: QFR-plot after Folk (1980) displaying the detrital composition including quartz (Q), feldspar (F), and rock fragments (R) of the samples for unbleached zones (red, N = 9) and bleached zones (blue, N = 11). Quartz is present in almost equal amounts in both sample types.

Generally, bleached zones are characterized by more blue-dyed intergranular pore space than is visible in unbleached areas (Fig. 2.5a, b). Intragranular macropores larger than the average grain size occur only in bleached zones (Fig. 2.5c). They are identified by their remaining dust rim (Fig. 2.5c), illite fibers surrounding the resulting pore space, and meshwork illite partially filling such intragranular macropores (Fig. 2.5f, h). Both, unbleached and bleached zones, are affected by intragranular dissolution with on average 0.4 % (max. 1 %, min. 0 %) and 1 % (max. 4 %, min. 0 %), respectively (Fig. 2.5c).

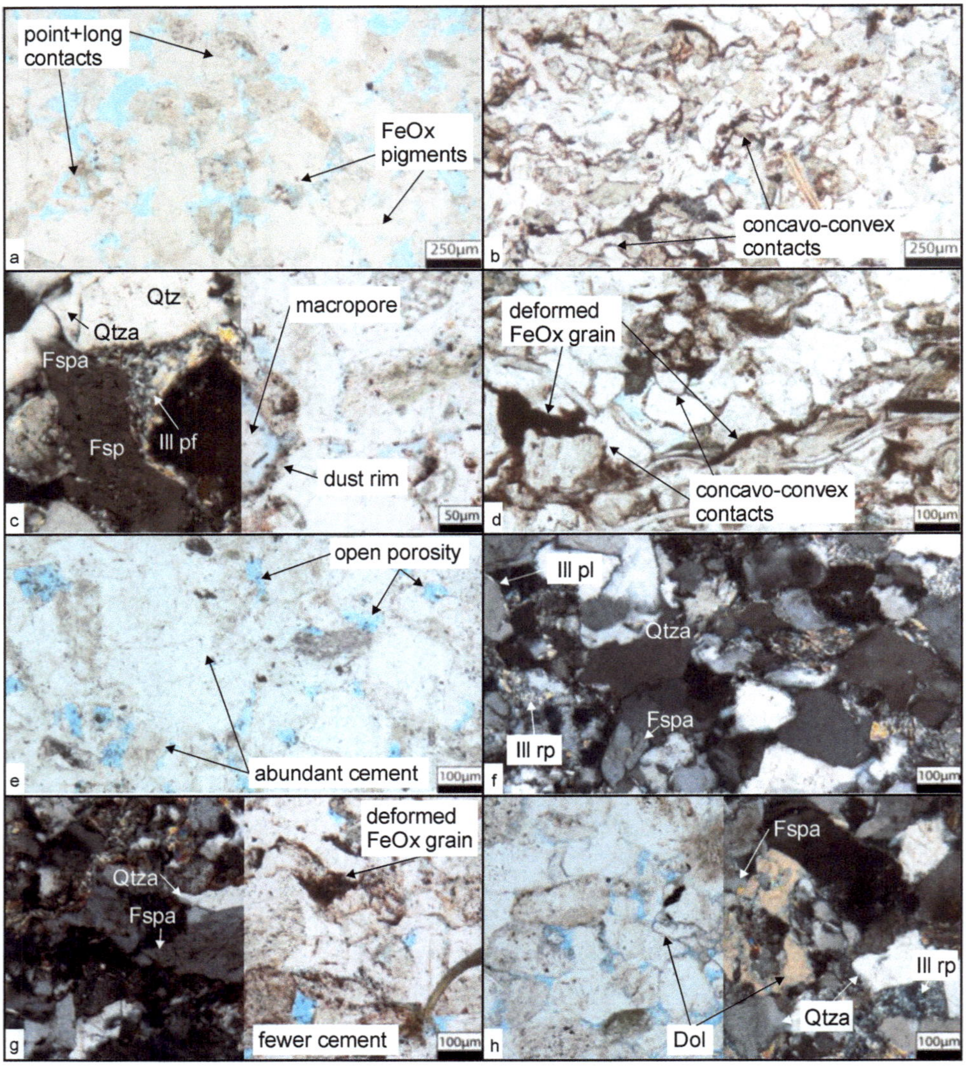

Figure 2.5: Figure caption on the next page.

Figure 2.5: Overview of mineral assemblage in bleached and unbleached zones. a) Bleached zone with point and long contacts as well as minor iron oxide (FeOx)-pigmented dust rims and with overall larger grain size (RB-28). b) Unbleached zone with concavo-convex grain contacts, abundant deformed FeOx-rich detritus, smaller grain sizes, and less pore space (RB-28). c) Intragranular macropore with remaining pigmented dust rim in bleached zone surrounded by pore-filling illite (Ill pf, plain-polarized light/ppl = right, cross-polarized light/xpl = left). Also visible are authigenic quartz (Qtza) limited by euhedral K-feldspar overgrowths (Fspa) on detrital grains (Qtz, Fsp, RB-17). d) Close-up of opaque and strongly deformed iron oxide grains and concavo-convex contacts in an unbleached zone (RB-24). e) Bleached zone with abundant pore-filling cements and open pores in ppl (RB-28). f) Same pore-filling cements can be identified as K-feldspar and quartz cements in xpl (Fspa, Qtza) with quartz growth stopping at euhedral K-feldspar cements. Pore-lining illite and illite replacing feldspars or rock fragments are observed (Ill pl, Ill rp, RB-28). g) Unbleached zone with less K-feldspar and syntaxial quartz cements (Fspa, Qtza), the latter stopping at the K-feldspar overgrowth (RB-13). Abundant deformed iron oxide grains are present (ppl = right, xpl = left). h) Authigenic dolomite (Dol) growing in the pore space and on detrital grains as well as cement surfaces of quartz or K-feldspar (Qtza, Fspa) and displaying a sometimes pitted appearance in ppl (left image). Illite replaces detrital grains (RB-28, ppl = left, xpl = left).

Grain contacts are primarily concavo-convex contacts in unbleached zones (Fig. 2.5b, d, 2.6). On average 56.9 % of concavo-convex, 3.7 % sutured, 37.5 % long contacts are counted and the remaining 1.9 % are point contacts and floating grains (Tab. A.1). A different trend is observed in bleached zones: Only 23.1 % of the contacts are concavo-convex, 0.3 % sutured, 47.3 % long, and 24.2 % point contacts as well as 5.1% floating grains (Tab. A.1). In all unbleached zones more than half of the grain contacts is contributed by concavo-convex and sutured contacts (51 to 77 %), while in bleached zones concavo-convex and sutured contacts make up 15 to 33 % (Fig. 2.6). The contact strength after Füchtbauer (1967) ranges for unbleached zones from 2.5 to 2.8 with an average value of 2.6. Bleached zones exhibit contact strengths from 1.9 to 2.2 with an average of 2.0 (Tab. A.1).

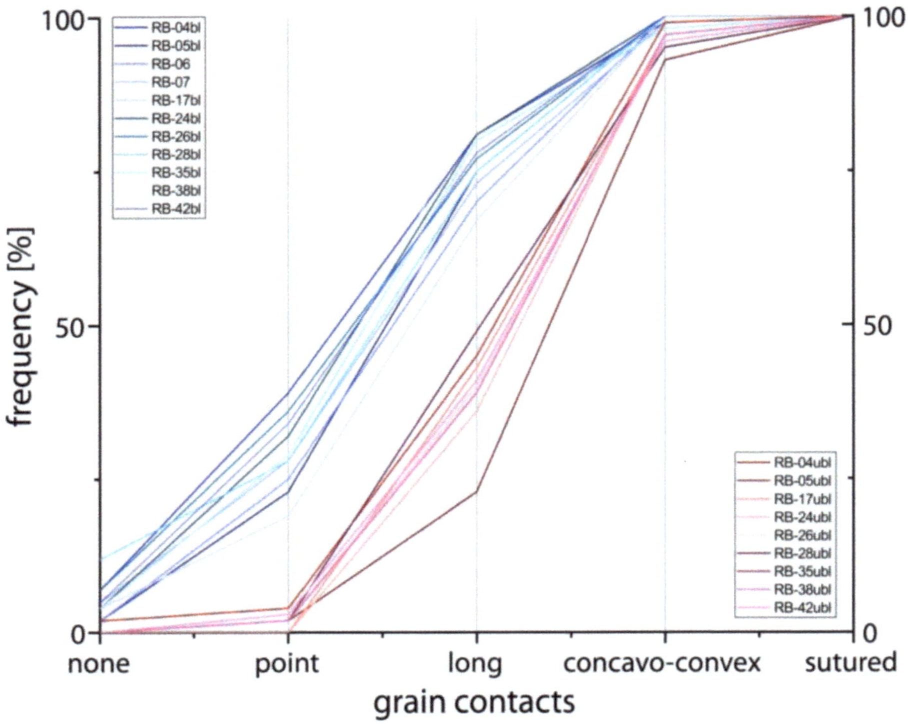

Figure 2.6: Cumulative curve of measured grain contacts for unbleached (red) and bleached (blue) zones in all partially and fully bleached samples. The outline indicates the coherently different behaviour of unbleached and bleached zones with a much higher frequency of point contacts in bleached zones, contrasted by primarily concavo-convex contacts in unbleached zones.

Iron oxides are present with on average 17.3 % in unbleached zones, bleached zones yield on average 0.5 %. A qualitative assessment shows that iron oxides are present in bleached zones constituting dust rims as pigmentation along the grain outline and underneath authigenic cements (Fig. 2.5a, c).

Quartz cements constitute 12 % in unbleached and 18 % in bleached parts (Fig. 2.5c, f, g, h, 2.7a, b). Similarly, the illite content in bleached zones (avg. 11%) exceeds that in unbleached ones (avg. 8 %, Fig. 2.5c, f, 2.7a, c). K-feldspar cement represents 7 % in the bleached zones whereas unbleached zones contain on average 4 % of K-feldspar cement (Fig. 2.5c, f, g, h, 2.7a, d). Other authigenic phases are dolomite (unbleached: avg. 0.1 %, bleached: avg. 0.5 %, Fig. 2.5h, 2.7a, e), which may appear pore-filling, sometimes with euhedral crystal habit, and often a pitted surface, titanium oxide (unbleached: avg. 0.1 %, bleached: avg. 0.1 %), replacive illite in alkali feldspars (unbleached: avg. 1%, bleached:

avg. 2%, Fig. 2.5f, h), and accessory replacive kaolinite or chlorite that essentially do not vary between bleached and unbleached zones (Tab. A.1).

The arithmetic means of the main pore-filling authigenic cements quartz, illite, and K-feldspar exceed the mean values of the unbleached zones by 3 to 6 %, whereas the mineral content of dolomite is too small to deduce a relationship (Fig. 2.7a). Similarly, for most of the nine samples quartz and illite cement contents are lower in unbleached zones (Fig. 2.7b, c, e), while the K-feldspar cement content is constantly higher in bleached zones (Fig. 2.7d). Authigenic minerals account for 23 % of the unbleached and for 34 % of the bleached zones.

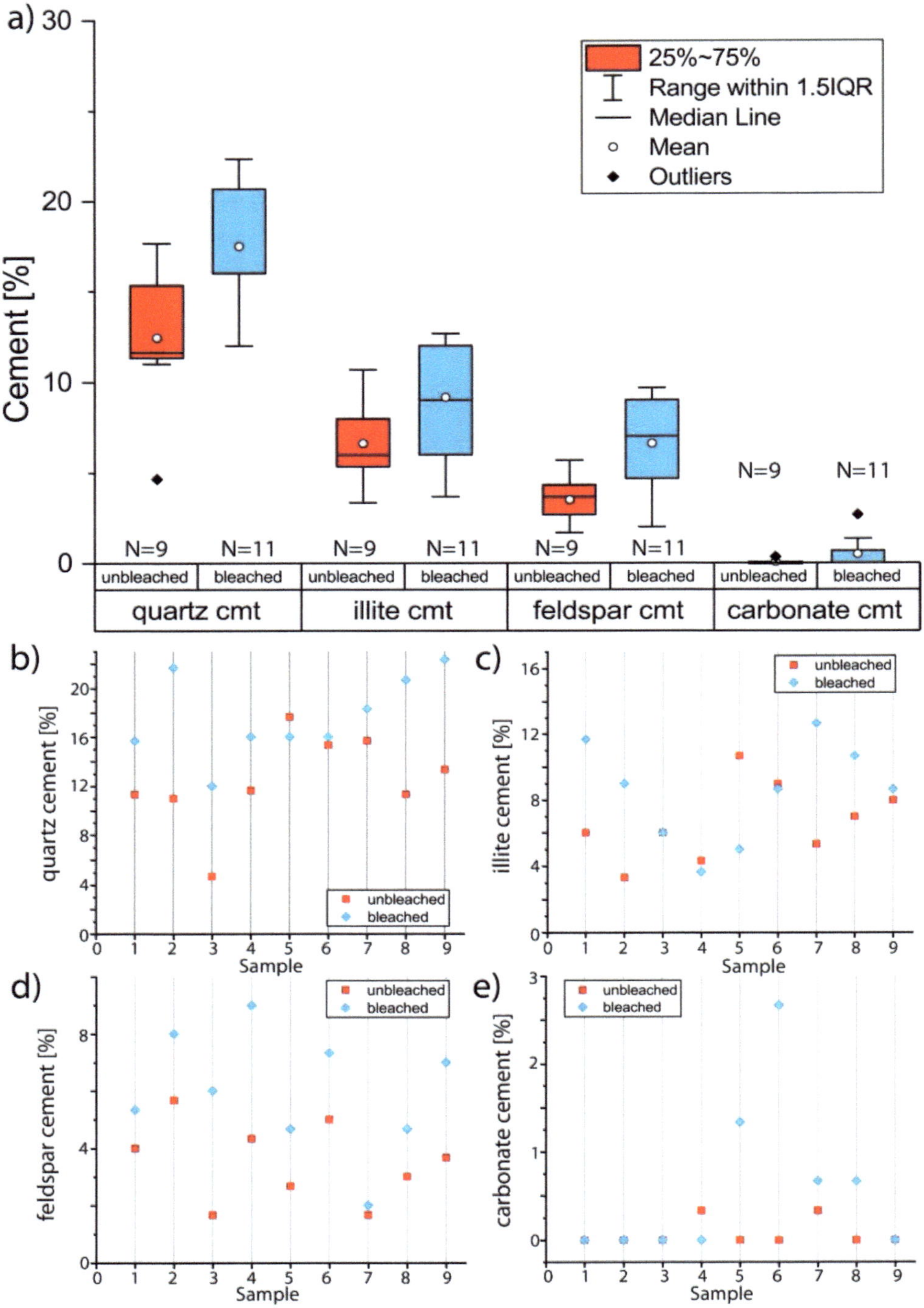

Figure 2.7: Figure caption on the next page.

Figure 2.7: a) Box-whisker plot of point counted cement contents (cmt) for all major authigenic phases. For quartz and K-feldspar cements, the 50th percentile does not overlap and for all types of cement the mean volume in bleached zones exceeds that in the unbleached zones. For carbonate (dolomite) cement, no quantitative proof can be given with regard to average volumes of < 1 %. b) Scatter plot for quartz cement. The cement content of eight samples is higher where bleaching has taken place (outlier RB-35). c) Scatter plot for illite cement: Cement occurrences are more frequent in bleached zones of five samples (outliers RB-24, RB-26, RB-28). d) Scatter plot for feldspar cements: All samples contain more K-feldspar cement in bleached zones. e) Scatter plot for dolomite cements: No quantitative proof can be given for a preferential precipitation of dolomite cement with regard to small average volumes of < 1 % and dolomite only being present in five of nine samples.

Point counted porosities for unbleached zones range from 7 to 15 % (avg. 11 %), while bleached zones exhibit increased point counted porosities of 11 to 18 % (avg. 14 %) (Fig. 2.8a). Values for all completely unbleached samples range from 9 to 17 % (avg. 12 %, Fig. 2.8a). The trend towards higher porosities in bleached zones is also reflected by each sample (Fig. 2.8b). Pycnometer analyses show porosities of 11 to 19 % (avg. 16 %) (Schmidt et al., 2020b). The average offset between petrophysical and point counted porosity accounts for +/-3 % (Tab. A.1).

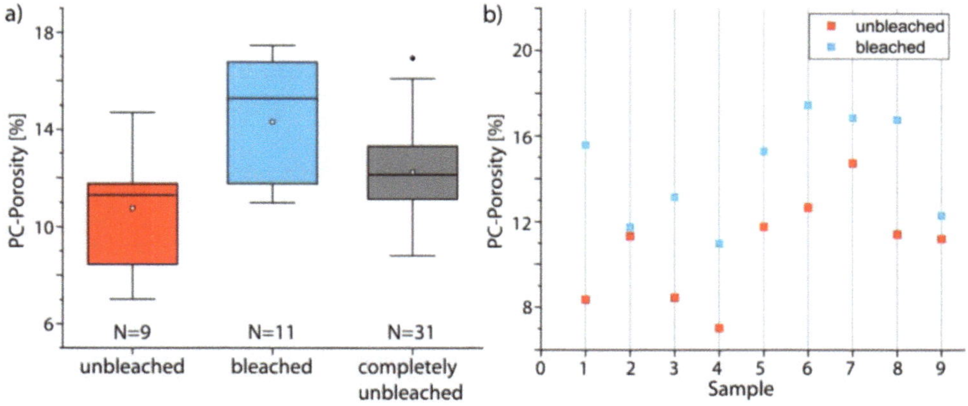

Figure 2.8: a) Box-whisker plot of point count-porosities (PC-Porosity) plotted for the three types of material: unbleached zones (red), bleached zones (blue), and completely unbleached samples (grey). Bleached zones contain 3 % more pore space (avg. 14 %) than unbleached rock (avg. 11 %). Completely unbleached samples cover most of the porosity range of the unbleached zones in partially bleached samples but do not exceed the mean porosity of bleached zones. b) The scatter plot for the nine partially bleached samples illustrates that bleaching in samples from the study area generally coincides with larger pore space.

Compaction

The intergranular volume (IGV) of unbleached zones accounts for on average 27 %, while bleached zones have an IGV of 39 % (Fig. 2.9). The average IGV-value of 33 % of completely unbleached samples (max. 38 %, min. 25 %) matches the IGV of unbleached zones (max. 34 %, min. 14 %) and overlaps with lower values of bleached zones (max. 43 %, min. 27 %, Fig. 2.9). The outlier exhibiting the lowest IGV among the unbleached zones is a sample containing rip-up clasts (RB-17).

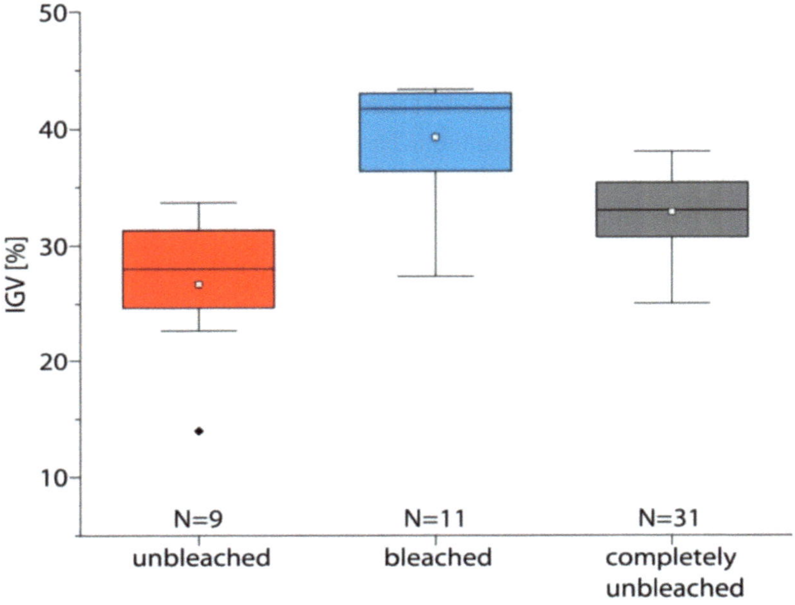

Figure 2.9: Box-whisker plot of the IGV plotted for the three types of samples: Unbleached zones (red), bleached zones (blue), and completely unbleached samples (grey). Results for bleached and unbleached zones plot in distinctly different ranges. The completely unbleached samples have an average IGV of 33 % and match with the results of unbleached zones. The lower range of IGV-values of bleached zones overlaps with the IGV of completely unbleached samples.

Considering compaction, unbleached zones show a compactional porosity loss (COPL) ranging from 17 to 36 % (avg. 25 %) and a cementational porosity loss (CEPL) ranging from 8 to 26 % (avg. 17 %, Fig. 2.10). In bleached zones the COPL-values range from 3 to 24 % (avg. 9 %), while CEPL-values range from 18 to 38 % (avg. 31 %, Fig. 2.10). The calculations after Lundegard (1992) suggest that almost all unbleached zones are dominated by compaction with the exception of two samples, whereas all bleached zones (except one sample) lose pore space primarily because of cementation (Fig. 2.10).

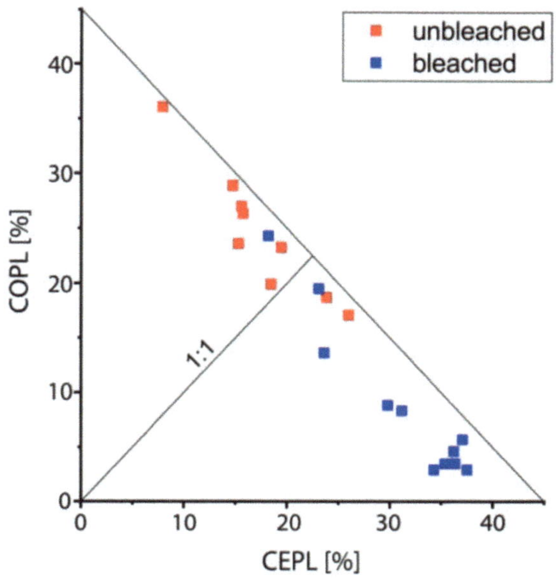

Figure 2.10: Porosity loss by compaction and cementation for all samples. Compactional porosity loss (COPL) is plotted against cementational porosity loss (CEPL) after Lundegard (1992). Results from the unbleached zones plot mainly in the upper left triangle representing sample material that has lost porosity primarily due to compaction (seven of nine samples). Results from bleached zones mainly plot in the lower right triangle indicating a reduction in porosity that is dominantly caused by cementation (ten of eleven samples).

2.6 Discussion

2.6.1 Macroscopic description, texture, and structure

Several studies present lamina-bound bleaching of red sandstones (Eichhubl et al., 2004; Nover et al., 2013; Pearce et al., 2011; Pudlo et al., 2012; Wendler et al., 2012). Discontinuous bleached spots of round to elliptical shape along a lamina indicate that bleaching is an authigenic process, likely unrelated to fluid flow, affecting an otherwise red deposit (Eichhubl et al., 2004). Stratigraphic constraints for bleaching (e.g. Eichhubl et al., 2004) are not visible in the studied 10 m thick sandstone beds. Thus, bleaching can be described as occurring intraformational within the associated sandstone beds. All bleached zones show irregular and diffuse transitions across laminae (Fig. 2.2b, c). Additionally, pore-lining remnants of hematite dust rims, also underneath quartz or K-feldspar overgrowths (Fig. 2.5a, c), are present in bleached zones, which are known from the Lower and Middle Permian or Lower Triassic (Busch et al., 2017; Kunkel et al., 2018; Schöner and Gaupp,

2005). Thus, bleaching removed hematite after its formation and after the associated reddening of the bulk rock (Walker, 1967b, 1975).

Grain size analyses reveal that bleached zones contain larger grain sizes (avg. 173 μm) than the unbleached zones (avg. 146 μm). The latter matches well with the bulk sample analysis of the same study area with an average grain size of 148 μm. Although differences are relatively small (on average 27 μm, Fig. 2.3a) in the very fine- to fine-grained sandstones, this effect is shown for every partially bleached sample (Fig. 2.3b). This result is supported by other authors that have qualitatively noted more coarse-grained lithologies where bleaching took place (Chu et al., 2015; Eichhubl et al., 2004; Kunkel et al., 2018; Pudlo et al., 2012; Taylor and Barker, 2002). Incorporating the thickness of a dust rim of 3 to 4 μm can also not explain the observed difference in grain diameters. Thus, grain size positively correlates with bleaching in the study area.

Better sorting is also reported to facilitate bleaching (Beard and Weyl, 1973; Pudlo et al., 2012) and may be supported by our data (Tab. A.1).

2.6.2 Petrography

Detrital composition in bleached and unbleached samples

The QFR-plot for bleached and unbleached zones in sandstones displays various rock compositions (Fig. 2.4, Tab. A.1). The average composition regarding quartz, feldspar, and rock fragments varies insignificantly by at maximum 3 %. Equal amounts of quartz grains indicate a similar transport distance of the detritus and thus maturity of the unbleached and bleached zones. This also negates that bleached laminae are associated with depositional processes.

Authigenic composition in bleached and unbleached samples

All major cement phases like quartz, illite, feldspar, and to a minor extent dolomite occur more frequently in bleached zones (Fig. 2.7). Similarly, other studies report more pronounced illitization (Chu et al., 2015; Pudlo et al., 2012; Schöner and Gaupp, 2005), more pronounced quartz cementation (Chu et al., 2015; Schöner and Gaupp, 2005) as well as more iron-rich carbonate and chlorite cements in bleached rocks (Beyer, 2015; Kunkel et al., 2018; Pudlo et al., 2012; Schöner and Gaupp, 2005). Such observations are contradicted by some authors reporting major quartz cementation in unbleached Lower Triassic rocks (Reinecker et al., 2015).

The petrographic analyses (Fig. 2.5a, e, f) illustrate that abundant pore space in the bleached zones was available for higher quartz and K-feldspar cementation. In bleached zones compaction has a minor effect on the morphology of grain contacts with more than one quarter being point contacts and floating grains (24.2 % and 5.1 %, respectively) as opposed to 23.5 % of compaction-related concavo-convex and sutured contacts (Fig. 2.6). In unbleached zones more than half of the grain contacts are compaction-related with 56.9 % concavo-convex and 3.7 % sutured contacts. Hence, compaction in bleached zones must have been prevented by an early, framework stabilizing cement phase in order to preserve uncompacted grain contacts and more pore space for quartz cementation during burial diagenesis. This is also supported by the higher calculated contact strength for bleached zones (Tab. A.1). The larger grain sizes may have, with higher permeabilities, facilitated more transport of available ions in the fluid for a more pronounced early diagenetic cementation along those laminae (Wilson et al., 2008).

A qualitative assessment yields that dolomite cements occur predominantly in the bleached laminae, which is not clearly visible from quantitative data (Fig. 2.7e). Fragmented, pore-filling dolomites have been linked to early diagenetic formation (Worden et al., 2018; Worden and Burley, 2003), whereas sometimes euhedral, pore-filling dolomite growth on quartz and K-feldspar cement is considered to be a burial to uplift diagenetic process (Gaupp et al., 1993). The pitted appearance of the dolomite cement patches can be assigned to uplift diagenetic dissolution by meteoric waters (Fig. 2.5h) (Worden et al., 2018). Staining did not reveal any significant iron content for the dolomites, nevertheless they could have incorporated minor amounts of iron after bleaching. To determine whether the iron left the system or was incorporated into the mineral assemblage, further analyses are needed as also chlorites replacing e.g. micas are only rarely observed (< 1.3 %).

Based on petrographic observations some dolomite and euhedral K-feldspar cements predate quartz cements and predate mechanical compaction (Fig. 2.5c, g, h). Thus, they formed during early diagenesis and stabilized the grain framework (e.g. Gaupp and Okkerman, 2011). Their stabilizing effect inhibited compaction, preserving intergranular volume to be cemented by later pore-filling phases. As some iron oxide pigments are still recognizable in bleached zones underneath syntaxial quartz- and also K-feldspar cements as well as underneath some of the dolomite cement (Fig. 2.5a), it can be assumed that bleaching and thus the removal of most iron oxide phases took place during this period of feldspar cementation and mostly prior to syntaxial quartz cementation.

Intragranular macropores are restricted to bleached zones. Possible detrital grains, which might have been dissolved to form these intragranular macropores, are K-feldspar grains or rock fragments, which can be observed as sometimes illitized remnants or voids (Fig. 2.5c, f, h). Schöner and Gaupp (2005) have linked macropores to the dissolution of feldspars and lithoclasts, while Kunkel et al. (2018) also proposed the leaching of early cements, e.g. sulfate and carbonate as well as feldspar. Feldspar dissolution constitutes a source for potassium ions as well as relatively immobile aluminum, allowing for a proximal precipitation of pore-filling illite (Bjørlykke, 2014; Hayes and Boles, 1992; Thyne et al., 2001). Feldspars thus constitute the most likely origin for macropores with illitization (Fig. 2.5f, g).

Two types of burial diagenetic illite are more prominent in the bleached zones (Tab. A.1). Dense, primarily pore-filling illite is occasionally present around pigmented hematite dust rims surrounding intragranular macropores (Fig. 2.5c), while meshwork illite locally also fills intragranular macropores (Fig. 2.5f, h). In order to preserve intragranular macropores, they must have formed after mechanical compaction. Similarly, illitization of feldspar grains, resulting in intragranular illite meshwork, likely postdates intragranular macropore formation and compaction (Molenaar and Felder, 2018). This process added secondary intragranular porosity (Fig. 2.5f, h). Illite replacements are associated with temperatures of commonly above 100°C and thus burial diagenesis (Bjørkum et al., 1993; Lander and Bonnell, 2010; Morad et al., 2010) although precipitation may also start at lower temperatures (Bauer et al., 1998; Bauer et al., 2000; Weibel, 1999; Wilkinson et al., 2006; Ziegler, 2006). Observations of a pronounced illitization in bleached samples also correspond with other studies (Chu et al., 2015; Gaupp and Schöner, 2008; Pudlo et al., 2012; Schöner and Gaupp, 2005) and indicate a burial diagenetic timing for the bleaching process.

The petrographic assessment yields 3 % more porosity in bleached zones (Fig. 2.8a), which is consistent with several publications (Gaupp and Schöner, 2008; Kunkel et al., 2018; Pudlo et al., 2012; Reinecker et al., 2015; Schöner and Gaupp, 2005; Wendler et al., 2012) and also with observations in petrographic images (Fig. 2.5a, b). However, the contribution of dissolution and resulting intragranular macropores contributes with only 0.4 and 1.1 % to porosity in unbleached and bleached zones, respectively (Tab. A.1). Higher porosity and permeability are causally linked with larger grain size as was shown for different grain size classes like fine-grained and medium-grained (Bloch, 1991; Chilingar, 1964). For point counted porosities this correlation exists also at small-scale grain size changes as they occur between unbleached and bleached zones of this study (Fig. 2.3). Similarly, sorting is described to positively correlate with porosity and permeability (Beard and Weyl, 1973), an effect that might also be present in this data set (Tab. A.1).

2.6.3 Compaction

Unbleached laminae are more compacted than bleached laminae, reflected by grain contacts and lower intergranular volume (avg. 27 %, max. 34 %, min. 14 %) than in bleached laminae (avg. 39 %, max. 43 %, min. 27 %). The grain contacts in unbleached areas (dominated by concavo-convex and sutured contacts with 56.9 % and 3.7 %, respectively) imply stronger pressure dissolution (chemical compaction) at grain contacts often covered by an illitic clay mineral. This reduction in intergranular volume results in smaller porosities and presumably permeabilities than in the bleached layers. The bleached layers contain more abundant K-feldspar and quartz cements reflected by 29.3 % point contacts and floating grains and almost no sutured contacts. All of which can potentially preserve higher intergranular volumes via framework stabilization.

Values for porosity loss by compaction versus cementation are distinctly different for unbleached and bleached zones. The COPL-CEPL plot after (Lundegard, 1992) supports the optical assessment in that unbleached samples show a more pronounced compactional porosity loss than the bleached samples (Fig. 2.10). The presence of early syntaxial K-feldspar cements in the now bleached areas preserved the intergranular volume and preserved porosity and permeability. The bleaching fluids then discolored the more coarse-grained laminae with higher porosity/permeability and higher fluid volumes to sufficiently react with the surrounding minerals.

2.6.4 Deriving possible bleaching processes

Bleaching has been encountered in close paragenesis with bitumen stains and thus organic matter, forming diagenetic clay mineral phases and removing hematite in a variety of case studies (Beitler et al., 2003; Gaupp and Schöner, 2008; Hilse et al., 2010; Pudlo et al., 2012; Schöner and Gaupp, 2005). Kunkel et al. (2018) describe bleached lithologies as a special diagenesis type and link its formation to meteoric fluids. Other studies analyzed stable carbon isotopy of calcites associated with bleaching and interpreted a volcanic or meteoric source (Kasch et al., 2010; Wendler et al., 2012). The absence of bitumen stains in samples from the study area and a lack of evidence for hydrocarbon-bearing underlying strata in the region suggests that no organic matter sourcing e.g. organic acids or methane was involved in partial bleaching of the red lithology (Beitler et al., 2003; Eichhubl et al., 2004; Gaupp and Schöner, 2008; Okrusch et al., 2011). The absence of organic matter also affects bacterial and thermochemical sulfate reduction, which can be excluded as major bleaching mechanism (Machel, 2001). The presence of volcanic dykes in the Spessart region may point toward a volcanic source of reducing fluids as it was proposed for Buntsandstein deposits in the the Palatinate Forest (Baaske, 1999; Peschel, 2013). But as dykes are assigned to the Upper Cretaceous to Tertiary (Okrusch et al., 2011), they can hardly coincide with an early burial diagenetic bleaching process as it is proposed in this study (Fig. 2.11). Pedogenesis, also known to cause a color alteration or color mottling in Lower Triassic deposits, is ruled out as a cause, since characteristic features like rootlets, nodular carbonate cementation, or calcrete formation have not been encountered in the outcrop (Molenaar, 1984; Ortlam, 1974).

Reducing brines from underlying salt are considered equally unlikely as cause for the bleaching, because the study area is situated at the former Triassic basin margin and underlying evaporitic successions of the Upper Permian Zechstein Group are unknown (Okrusch et al., 2011). Furthermore, no pyrite neoformation was registered during point counting, which would have indicated bleaching by H2S (Parry et al., 2004). However, reducing fluids from the overlying Muschelkalk sea with local evaporation pans (Okrusch et al., 2011) may have caused lamina-bound to pervasive bleaching in the sampled outcrop.

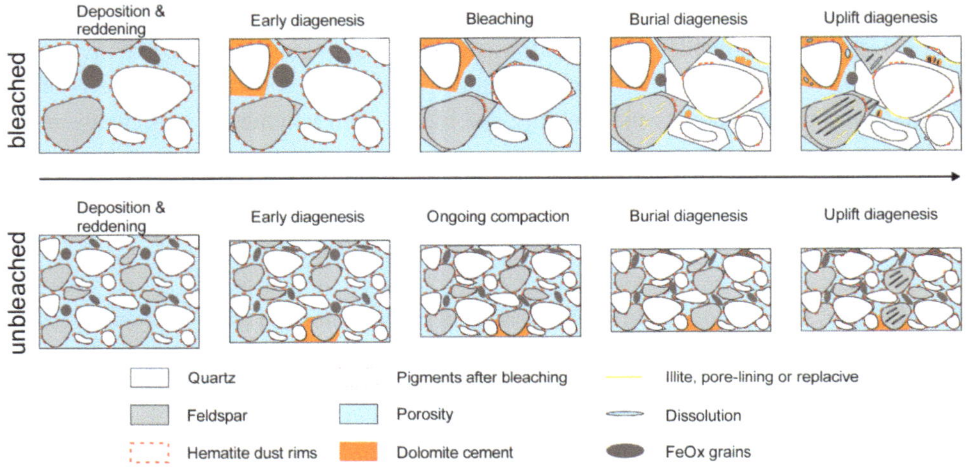

Figure 2.11: Schematic overview of the bleaching process and diagenetic alterations. The upper row displays the temporal diagenetic alteration in the more coarse-grained laminae with cementational porosity loss and carbonate and feldspar dissolution during uplift. The lower row shows the alteration in the finer-grained unbleached laminae with compactional porosity loss, less pore space, and more chemical compaction at sutured and concavo-convex grain contacts (lower row).

2.7 Conclusion

Bleached zones exhibit a grain size larger by 27 µm and to a minor extent better sorting. The present-day mineralogy is similar but porosities are on average 3 % higher in bleached laminae and thus reflect better reservoir quality. It can be assumed that initially better porosity and permeability enabled the fluid flow along specific laminae and their bleaching.

Compaction is higher in unbleached zones featuring primarily concavo-convex and sutured contacts (61 % vs. 24 % in bleached zones) and a lower IGV (27 % vs. 39 % in bleached zones).

Early diagenetic cement phases (dolomite and K-feldspar cement) stabilized the grain framework during compaction in bleached zones, preserving initially higher porosities and thus pathways that enabled bleaching. Bleached zones show a larger intergranular volume available to burial diagenetic quartz cementation and exhibit 11 % more diagenetic minerals. Intragranular macropores are limited to bleached laminae and form secondary porosity. They are occasionally filled with illite fibers and formed post-compaction by the dissolution of feldspars and rock fragments. The more coarse-grained and more quartz- and feldspar-cemented bleached laminae exhibit higher IGV and point counted porosity. Those laminae also allowed subsequent leaching of detrital grains (rock fragments and feldspars) and minor amounts of carbonate cements during uplift diagenesis due to their likely higher initial permeability. In contrast, finer-grained laminae show more compaction reducing available intergranular volumes for cementation and reducing porosity and permeability and thus reservoir quality.

Overall, bleaching is promoted in laminae with larger initial grain size. Bleaching is thus the consequence and not the cause of primarily better permeability along single laminae.

3 Lateral variations of detrital, authigenic, and petrophysical properties in an outcrop analog of the fluvial Plattensandstein, Lower Triassic, S-Germany

3.1 Abstract

The lateral heterogeneity of reservoir rocks is important to assess storage potential and permeability. This case study focusses on a Lower Triassic (upper Olenekian) outcrop of fluvial sandstones of the Plattensandstein Member (Röt Formation) near Röttbach in the southern part of the Germanic basin, Southern Germany. The Buntsandstein Group is considered a suitable reservoir for thermal storage or exploration in Southern Germany. In the Röttbach Quarry, the > 10 m thick lithic arkoses are exposed over 35 m laterally in three adjacent walls and consist of two units separated by an erosive surface. They are covered by several meters of the Lower Röt Claystone Member. Using sedimentological logging, spectral gamma ray-, porosity-, and permeability measurements as well as petrographic analyses calculating compaction parameters, lateral reservoir quality differences are studied.

Sedimentary structures in the multi-storey channels suggest a point bar deposit. The migration direction of the channels correlates with deteriorating reservoir quality. This reduction in porosity and permeability correlates with an increased detrital mica and authigenic illite content, the occurrence of rip-up clasts and a lateral increase of gamma ray signatures by 19 API (avg. 5.0 % K, 2.7 ppm U, 11.9 ppm Th). The occurrence of larger amounts of mica and clay minerals is linked to decreasing flow velocities throughout the evolution of a meandering system. They enhance compaction and reduce available intergranular pore space.

Permeability laterally deteriorates by up to three orders of magnitude (48 to 0.02 mD, avg. 9 mD), while measured porosities show minor variability (11 to 19 %, avg. 16 %). The deterioration of reservoir quality is related to compactional (COPL 11 to 28 %, avg. 18 %)

and cementational porosity loss (CEPL 13 to 29 %, avg. 22 %) and is mainly controlled by detrital mica and authigenic illite.

3.2 Introduction

Reservoir sandstones of fluvial origin exhibit heterogeneities on multiple scales that are difficult to detect when relying solely on vertical profiles or 1D-well information (Bridge, 2006; Miall, 1977a, 1988a; Miall and Turner-Peterson, 1989; Pranter et al., 2007; Taylor and Ritts, 2004). Porosity and permeability and thus reservoir quality can be controlled by e.g. the overall depositional system, moving down in hierarchy to the fluvial style and the sorting, size, and composition of individual sand grains (Bridge, 2006; Miall, 1988a, 2014; Morad et al., 2010). The categories micro-, meso-, macro-, and megascopic heterogeneity have been introduced that cover characteristics observed on the grain scale, the lithofacies scale, the scale of the specific architectural element, and the scale of the over-all stratigraphy, respectively (Miall, 1996; Tyler and Finley, 1991). Often core or well data cannot provide the information necessary to identify architectural elements from lithofacies types represented in the sample material or image logs. Additionally, they lack information on lateral changes making it difficult to predict rock properties away from the core sample or well (Becker et al., 2017; Miall, 1977a, 1988a; Miall and Turner-Peterson, 1989). Outcrop analog studies offer a means to assess reservoir quality with respect to micro- up to macro-, sometimes even megascopic heterogeneity displayed in 2D or 3D along outcrop walls utilizing thin sections, plug samples, and macroscopic observations (Becker et al., 2017; Bridge, 2006; Miall, 1988a; Pranter et al., 2007; Trendell et al., 2012; Wachutka and Aigner, 2001).

Heterogeneity of bedsets may include the detrital composition, authigenic alteration, sedimentary structure, or texture (Pranter et al., 2007; Trendell et al., 2012). Volumes of quartz cement often correlate negatively with grain coating coverages on detrital grains (Heald and Larese, 1974). The extent of grain coating coverage at grain contacts was also recently related to the degree of chemical compaction (Monsees et al., 2020). Compaction is facilitated by both mechanical (grain rearrangement and deformation) and chemical compaction (pressure dissolution of e.g. quartz grains) (Bjørkum et al., 1993; Walderhaug, 1996). A higher amount of ductile clasts among the detrital framework grains (e.g. micas or shale rock fragments) enhances mechanical compaction. Ductile clasts may form a pseudomatrix as they undergo strong deformation during compaction (Becker et al., 2017; Bertier et al., 2008; Busch et al., 2019; Geyer, 2005; Henares et al., 2016a; Molenaar et al., 2015). In diagenetic systems with early cementation, mineral growth of authigenic phases such as feldspar and carbonate may stabilize the grain framework against mechanical compaction during subsidence and may further enhance reservoir quality when dissolved at depth (Canham et al., 1996; Henares et al., 2016b; Line et al., 2018; Lü et al., 2015; Paxton et al.,

2002; Scherer, 1987; Walderhaug, 1996, 2000). The distribution of detrital clay and mica minerals in the mineral framework causes a reduction of reservoir quality for similar rock types affected by the same subsidence history (Busch et al., 2020; Busch et al., 2017; Fischer et al., 2012; Gaupp et al., 1993; Lander et al., 2008; Morad et al., 2010; Walderhaug, 2000). During chemical compaction, mineral dissolution and precipitation impacted by increased contents of mica and clay minerals significantly change the pore volume and its connectivity (Beyer et al., 2014; Bjørkum et al., 1993; Bjørlykke, 2014; Gaupp et al., 1993; Lundegard, 1992; Monsees et al., 2020; Morad et al., 2010; Paxton et al., 2002). Authigenic phases such as feldspar-, carbonate- or quartz cements that reduce pore space mostly deteriorate reservoir quality. Authigenic clay minerals like kaolinite or illite reduce permeability and porosity during early and burial diagenesis (Bjørkum et al., 1993; Bjørlykke, 2014; Gaupp et al., 1993; Lander and Walderhaug, 1999; Morad et al., 2010; Taylor et al., 2010; Walderhaug, 1996, 2000).

Petrophysical properties may vary in fluvial deposits by several orders of magnitude. Doyle and Sweet (1995) measured porosities of 5.4 % to 30.4 % and permeability ranges over six orders of magnitude within a 15 x 35 m outcrop representing stacked meander channels. They proposed the lithofacies type as reservoir quality-controlling factor. For example, main vertical permeability-reducing baffles present as mud clast lags at the channel base and mud drapes on lateral accretion surfaces (Doyle and Sweet, 1995). Permeability variations over three orders of magnitude were also recognized on a scale of < 3 m for braided channel sandstone beds (McKinley et al., 2011), linked to the type of underlying erosional bounding surface and the successive deposition. Depositional parameters (grain size, sorting) controlled petrophysical properties above scour troughs, whereas in continuous channel fill deposits diagenetic cements reduce reservoir quality (McKinley et al., 2011). Equal variability has been observed in composite point bar deposits. Here, permeability reduction was linked to a cemented mud horizon connected to the floodplain generated during channel abandonment (Hartkamp-Bakker and Donselaar, 2009). Differences between fore- and bottomsets have been determined with bottomsets of lower permeabilities due to smaller grain size and poorer sorting (Hartkamp-Bakker and Donselaar, 2009). Becker et al. (2017) showed permeability changes from 0.001 to 0.1 mD on a larger scale over a horizontal profile of 150 m in a single fluvial sandstone coset related to cm-scale fining-upwards grading and internal erosion surfaces (Becker et al., 2017).

In this study, lateral variations in sandstone composition, porosity, and permeability of a macroscopically homogenous, fluvial Lower Triassic Buntsandstein deposit on the outcrop scale are studied. Petrophysical heterogeneities of the fluvial sandstones evolving with progressive lateral accretion at a point bar in a meandering river system are presented. The effect of meander evolution influencing the detrital and authigenic mineral contents is discussed. The main reservoir quality controls due to detrital and authigenic compositions and their effect on compaction are presented.

3.3 Geological setting

The outcrop in Röttbach, 5 km North of Wertheim, Lower Franconia, Southern Germany, features red fluvial sandstones of the Lower Triassic (late Olenekian) Plattensandstein (Röt Formation, Upper Buntsandstein) (Deutsche Stratigraphische Kommission, 2016; Menning, 2018; Röhling et al., 2018) (Fig. 3.1a, b). Those deposits are part of the southwestern marginal facies of the southern extension of the Germanic Basin, where the Buntsandstein extends over approximately 600 km from the Burgundy Basin to the Center of the Germanic Basin (Backhaus, 1974; Freudenberger et al., 2013; Hagdorn et al., 2009; Okrusch et al., 2011; Ziegler, 1990) (Fig. 3.1a). Of the 530 m thick Buntsandstein succession in Lower Franconia, 25 to 35 m are Plattensandstein in the study area (Bock et al., 2005; Freudenberger, 1990; Geyer, 2002). The Plattensandstein contains intercalations of claystone, which account for approximately 10 to 20 % of the overall rock unit in the region (Freudenberger, 1990). The epicontinental basin formed by minor thermal subsidence rates and ongoing sedimentation during the Triassic (Meschede and Warr, 2019; Röhling and Lepper, 2013). Sedimentation was mainly controlled by local paleotopography (Hagdorn et al., 2009; Röhling and Lepper, 2013; Röhling et al., 2018; Schröder, 1982). The detritus was shed from the Vindelician High from the southeast and the Ardennes and Gallic Massif in the West while the overall transport took place to the northeast (Backhaus, 1974; Bock et al., 2005; Freudenberger et al., 2013; Geyer, 2002; Hagdorn et al., 2009; Meschede and Warr, 2019; Röhling and Lepper, 2013; Röhling et al., 2018; Schultz-Krutisch and Heller, 1985; Ziegler, 1982) (Fig. 3.1a).

The marginal facies developed from braided rivers in the Early and Middle Buntsandstein (Induan to early Olenekian) to meanders in the Upper Buntsandstein (late Olenekian) (Bock et al., 2005; Freudenberger et al., 2013; Geyer, 2002; Okrusch et al., 2011), the latter being exposed in the quarry. Corresponding channels of the Upper Buntsandstein form shallow incisions of up to a few meters in depth and meander laterally over a substantial width but at a relative constant position (Bock et al., 2005; Freudenberger et al., 2013; Geyer, 2002; Okrusch et al., 2011). The Plattensandstein itself represents meandering river systems shedding sediment in a low-relief area with occasional flooding events (Bock et al., 2005; Geyer, 2002). Plant and trace fossils document a mainly terres-trial setting (Geyer, 2002; Kirchner, 1927; Trusheim, 1937) but horizons of marine bivalves reaching into the Odenwald area imply a marine influence already in the Plattensandstein and in the subsequent Röt Claystones (Frank, 1937; Geyer, 2002; Martini, 1992).

The upper stratigraphic succession exposed in the quarry consists of Lower to Middle Triassic (late Olenekian to Anisian) Lower Röt Claystones (Upper Buntsandstein), which were deposited in a sabkha or playa environment (Aigner and Bachmann, 1992; Okrusch et al., 2011). Prevailing lithologies are silt- and claystones, which may feature salt or sulfate nodules (Meschede and Warr, 2019; Okrusch et al., 2011). This reflects the intermittent ingression of the Tethys Ocean from the South in the Upper Buntsandstein, at the transition from the Lower to Middle Triassic (Aigner and Bachmann, 1992; Backhaus, 1981; Freudenberger et al., 2013; Okrusch et al., 2011; Schröder, 1982; Ziegler, 1982). Warm, humid climates during the Tertiary could have led to intense weathering and pedogenesis in preceding formations (Böse et al., 2018).

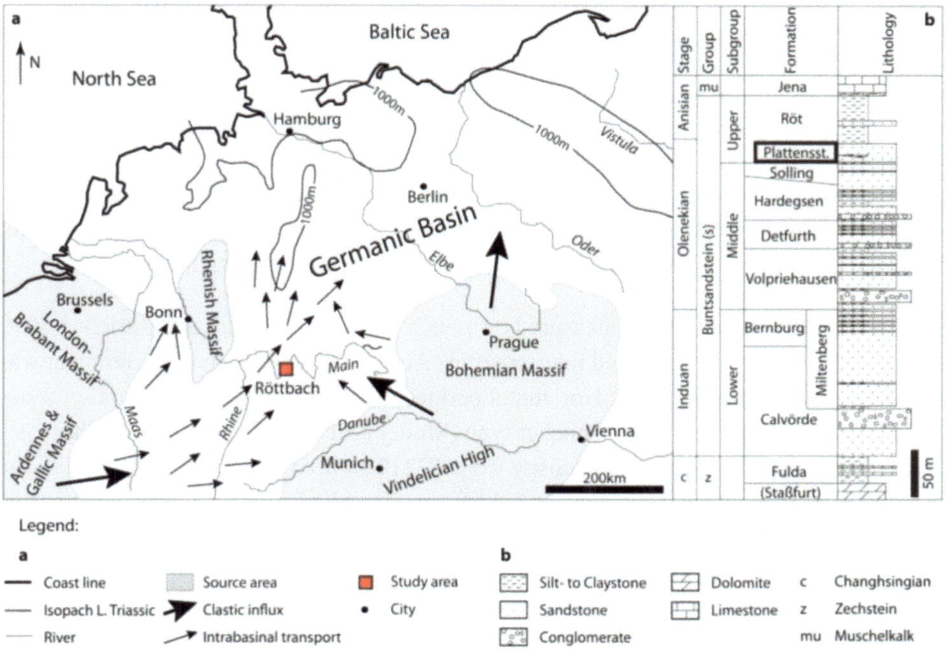

Figure 3.1: a) Location of the study area (red square) in the Germanic basin during the Lower Triassic. Sediment transport (indicated by arrows) from the Vindelician-, Gallic-, and Ardennes highs (after Geyer et al., 2011, Ziegler, 1990). Isopachs display the maximum sediment thicknesses for the Buntsandstein, Lower Triassic, in the wider region. b) Local stratigraphy: The Plattensandstein Member (black square) is present in the footwall of the Lower Röt Claystone Member, Röt Formation (Upper Buntsandstein Subgroup, upper Olenekian) after Okrusch et al. (2011) modified from Geyer (2002).

3.4 Materials and methods

Along three profiles of vertically stacked sandstone beds extending along 35 m, 1-inch-diameter plugs were drilled parallel to bedding (N = 36, Fig. 3.2a). The three sampled profiles originated from two beds separated by a subhorizontal erosional surface. Generally, the bedding was poorly visible in the active quarry due to the extraction method combining a rope-saw system with vertical drill holes. Profile 1 was the basal profile with distances between 1.60 and 2.50 m from the top contact to the overlying claystone, situated at a central position in the basal unit (N = 14, Fig. 3.2a). Profile 2 was sampled 0.30 – 0.45 m above profile 1 and within the same unit (N = 14). In its hanging wall profile 3 was located in the subsequent bed, 0.75 – 1.20 m above profile 1 (N = 8). In each profile two samples originated from the SE-wall (Orientation: 320/90°) and two from the NW-wall (Orientation: 126/82°), all others were extracted from the SW-quarry wall. The 24 samples from the SW-wall were plugs of one horizontal direction, whereas the remaining 12 samples from the SE- and NW-wall represented a second, sub-perpendicular horizontal direction. The produced one-inch plugs were sawed to lengths between 2.5 to 4.0 cm. Trim ends were used for thin section preparation.

Bounding surfaces of a fluvial origin were classified according to Miall (1988a, 1988b). Six orders were subdivided based on concordance of set bedding contacts, changes in direction of deposition, erosion, facies assemblages, and architectural elements (Miall, 1985), concordant sand sheet contacts, and contacts between stratigraphic units.

Porosity and permeability were measured on 1-inch plugs, which were oven-dried at 40°C for one week. The semi-automated micromeritics Accupyc II 1340 helium pycnometer was used for porosity measurements. Horizontal permeabilities were determined in a semi-automated isostatic flow cell of the Fancher type (Model Rieckmann) using single-phase gas flow. Compressed air (measured viscosity $\eta = 1.82*10^{-5} - 1.84*10^{-5}$ Pa*s) was used as the permeant while a confining pressure of 1.2 MPa was applied. Permeabilities were calculated from the flow rates according to Darcy's law and were Klinkenberg-corrected for slip flow (Klinkenberg, 1941; Rieckmann, 1970).

Trim ends were oven-dried at 40°C for a week and embedded in a blue-dyed epoxy and prepared to a thickness of 30 µm. A combined staining of Alizarin Red S and potassium ferricyanide in HCl was applied to aid in the identification of carbonate minerals. Iron-free dolomite shows no stain from the treatment with this solution (Dickson, 1966).

For grain size analyses the long axes of 100 detrital grains were measured for each thin section on a grid adjusted to the largest grain size (Busch et al., 2018). Grain size classes were assigned according to the Udden-Wentworth scale (Udden, 1914; Wentworth, 1922), sorting was classified after Folk and Ward (1957). Grain roundness was classified using comparative images after Füchtbauer (1988c).

Point counting was performed with 300 steps per sample on a grid adjusted to the maximum grain size using a semi-automated Pelcon point counter mounted on a Leitz Aristomet microscope. Results were plotted after Folk (1980). Errors on point counted mineral phases were determined via the half-width of the lower and upper 95 %-confidence bounds as proposed by Howarth (1998). Point count values were rounded to derive a lower and upper bound for each sample. Samples RB-17 and RB-19 contained rip-up clasts with sizes of several mm, which would have skewed the overall area-weighted results during point counting and were not considered for grain size assessments and point counting. All observed types of iron oxide (pore-lining and filling, grains, iron-rich pseudomatrix) were combined in a single class for point counting. As ductile deformation during compaction prohibited any clear distinction between authigenic and detrital iron oxide phases, the point counted class was assigned to the detrital components.

Optical porosity was generally lower than measured porosity since microporosity in clays could not be sufficiently detected in petrographic analyses. To account for microporosity, point counts for kaolinite, chlorite, and illite were multiplied with mean fractions of microporosities in those mineral phases after Hurst and Nadeau (1995) and the average of those values was adopted for undifferentiated clay minerals. The calculated microporosity and the point counted inter- and intragranular pore space were combined in the category "point counted porosity".

The intergranular volume (IGV) was calculated according to Paxton et al. (2002) to assess compaction based on point counting results. It comprises the intergranular pore space, intergranular cement volumes, and the depositional matrix (Eq. 3.1). By calculating the minus-cement porosity after Lundegard (1992), adding up optical porosity and intergranular cement volume, the amount of porosity loss due to compaction (COPL) or cementation (CEPL) could be determined following equations 3.2 and 3.3.

$$IGV = \phi_{intergr.} + V_{cement} + V_{matrix} \tag{3.1}$$

$$COPL = \phi_{initial} - \frac{[(100 - \phi_{initial}) \times IGV]}{100 - IGV} \tag{3.2}$$

$$CEPL = (\phi_{initial} - COPL) \times \frac{V_{cement}}{IGV} \tag{3.3}$$

Variables are the initial porosity ϕinitial, the present-day, point counted values for the intergranular porosity ϕintergr., the cement volume Vcement as well as the matrix volume Vmatrix for each sample. The initial porosity in a fluvial system was assumed to be 45 % after Wilson and McBride (1988).

Spectral gamma ray (SGR) was measured at 14 spots along profile 1 using the hand-held, auto-stabilizing spectrometer RS-230 BGO Super-SPEC. The device detects potassium, uranium, and thorium decays over a time span of 3 minutes and calculates element contents. Results are given in API units calculated after Rider and Kennedy (2011) (Eq. 3.4).

$$SGR(API) = 16.32API \times K(\%) + 8.09API \times U(ppm) + 3.93API \times Th(ppm) \tag{3.4}$$

The error of 3.4 % on the API-value was determined as the standard deviation of 10 measurements of the same sandstone sample.

3.5 Results

3.5.1 Sedimentological logging

The up to 10 m high outcrop is composed of two subhorizontal red sandstone beds with decimeter to two meter thick cosets (Fig. 3.2a). Macroscopically, only few sedimentary structures are recognizable in the outcrop and included in the sedimentological sketch and log (Fig. 3.2). Coset boundaries in the study area coincide with third-order bounding surfaces after Miall (1988a). Third-order surfaces erode underlying strata at a low angle and exhibit similar facies assemblages underneath and on top of the surface (no. 3, Fig. 3.2). Some sets within the cosets are pinching out to the NW and their outlines correspond to first-order bounding surfaces of equivalent bedforms (no. 1, Fig. 3.2). Tangential foresets of the SE-wall may reach angles of 20° (no. 1, Fig. 3.2a), whereas the base and top boundaries bracketing single cosets mainly exhibit a shallow or subhorizontal dip (no. 3, Fig. 3.2).

The basal, up to 1.8 m thick bed consisting of fine-grained sandstone (profiles 1 and 2) shows two channel-like concave-up surfaces in the SW-quarry wall of 2 m and 4 m width (no. 2, Fig. 3.2). They are classified as second-order surfaces because of an apparent change in flow direction. The same basal bed exposes more strongly inclined (up to 20°) foresets in the SE-wall. It is overlain by an up to 1.2 m thick bed, which consists of similar fine-grained sandstone. Both units are separated by an erosive surface (no. 3, Fig. 3.2). This third-order surface marks the upper limit of the basal coset with the overlying one exhibiting the same apparent dip direction. The upper bed displays few shallowly (< 10°) dipping surfaces on the NW-wall and some strongly inclined, tangential surfaces on the SE-wall (concave-up white lines, Fig. 3.2a). The SW-wall exposes two surfaces of very shallow apparent dip to the NW that behave asymptotic towards the base of the bed, continuing each for > 10 m (no. 1, Fig. 3.2). Those recurring first-order surfaces mirror the shallowly NW-dipping surfaces present in the basal coset. All of those surfaces outline sigmoidal sandstone units. The transition to the Röt Claystone Member in the hanging wall is constituted by a 0.5 m thick, weathered unit of silt- to sandstone at the SW-wall, which is separated from the underlying sandstone unit by a discrete horizontal, erosive surface truncating sedimentary structures of the footwall (no. 5, Fig. 3.2). The bounding surface was classified as fifth-order because it limits a major sand sheet deposit in its footwall (no. 5, Fig. 3.2). At the SE-wall another 0.4 m thick sandstone coset is present, separated from the underlying sandstone by another second-order bounding surface between the two cosets (no. 2,

Fig. 3.2a). The overall silty to sandy lithology is then overlain by the claystones of the Röt Claystone Member separated by a sixth-order bounding surface tracing mappable, stratigraphic units (no. 6, Fig. 3.2a).

Irregular, reddish-brown, elongated, angular rip-up clasts of up to 7.5 cm are observed in mined blocks and the exposed basal sandstone coset and appear predominantly in the NW-part (sketch, Fig. 3.2a). They consist of clayey to silty material and exceed the thin section-scale in length in two samples (RB-17, RB-19). Millimeter-scale mica flakes are ubiquitous on all bedding planes. No tropical weathering nor other types of paleosol can be identified in the outcrop.

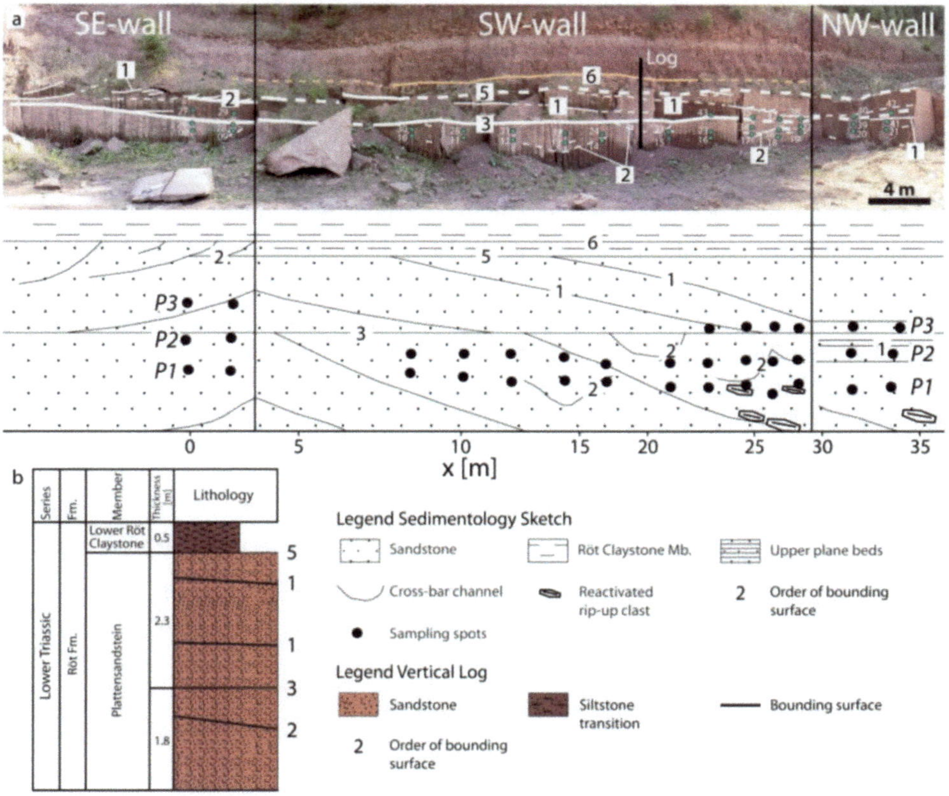

Figure 3.2: Figure caption on the next page.

Figure 3.2: a) Panorama photo of the studied quarry and the corresponding, vertically exaggerated, sedimentological sketch (orientation changes are indicated by the vertical black lines; dashed and solid white lines indicate interpreted structures). Three profiles (P1 to P3) were sampled along the SE-, SW-, and NW-wall of the active quarry (black dots). The SGR-profile was measured along P1. The orange line, 6, highlights the border between the Plattensandstein and the Lower Röt Claystone Member (both Röt Formation) in the hanging wall (s. Fig. 3.1b). Numbers indicate the order of the observed bounding surface after Miall (1988a). White lines follow sedimentary structures as coset boundaries, tangential cross-bed sets, and minor channels etc. and are drawn as dashed lines where poorly visible. b) Vertical log (location in a)) showing the stratigraphy, lithology, and visible sedimentary structures.

3.5.2 Petrography

Detrital composition

The samples are classified as lithic arkoses (N = 21), subarkoses (N = 8), feldspathic litharenites (N = 3), and arkoses (N = 2) (Fig. 3.3). The average quartz grain content is 35 % (max. 41 %, min. 23 %) (Tab. A.1). The average feldspar grain content is 8 % (max. 12 %, min. 6 %), while rock fragments amount to 6 % (max. 9 %, min. 3 %) (Tab. A.1). The observed lithology consists of very fine- (N = 8) to fine-grained (N = 26) sandstone with average grain sizes of 110 to 170 µm (avg. 150 µm). The sand grains are subangular to subrounded and moderately (N = 33) to well sorted (N = 1) (Tab. A.1). Grains are sometimes aligned, while micas always adjust their long axes parallel to bedding planes. From qualitative assessment, grain contacts are primarily concavo-convex contacts with only few sutured and point contacts indicating a well compacted sandstone.

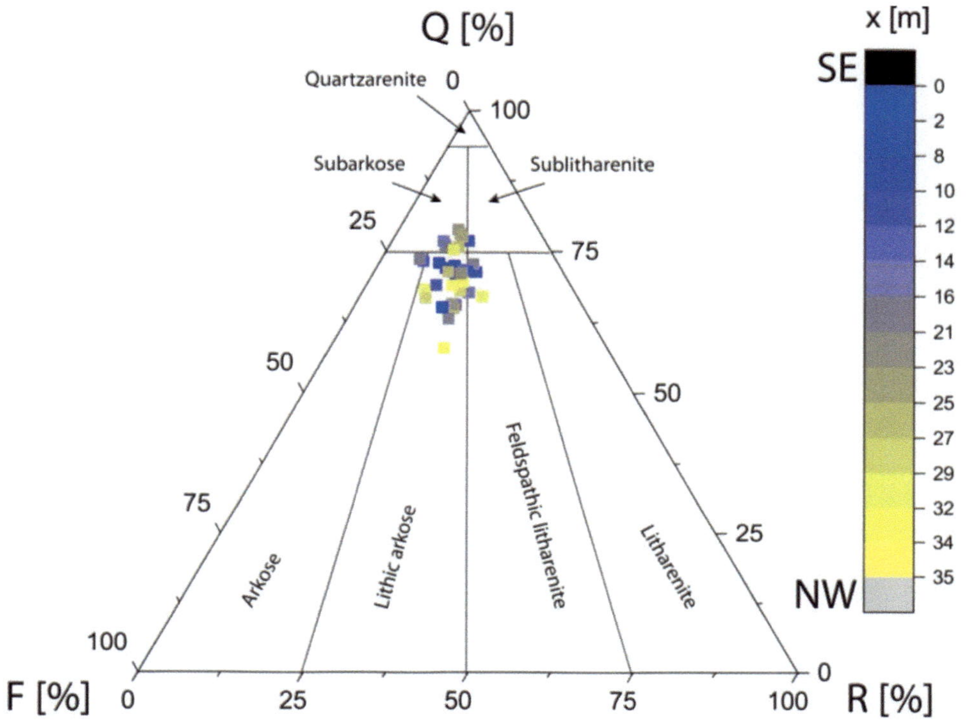

Figure 3.3: QFR-plot of the sandstone samples (N = 34) from the three profiles in the studied Plattensandstein outcrop, Röt Formation (blue = SE-part of the outcrop, yellow = NW-part of the outcrop).

The major detrital component quartz is present as mono- and polycrystalline grains frequently displaying undulose extinction. Feldspars are mostly K-feldspars, plagioclases are rarely preserved and not encountered during point counting. K-feldspars are often partially dissolved. Rock fragments include sandstone, plutonic, and metamorphic rock fragments, phyllite, chert, ductile siltstone, claystone, and shale fragments (Fig. 3.4a) and may show partial dissolution. The intragranular pore space in feldspars and rock fragments only contributes up to 3.3 % (avg. < 1 %) and does not significantly increase the overall point counted porosity (avg. 13 %, max. 17 %, min. 9 %) (Tab. A.1).

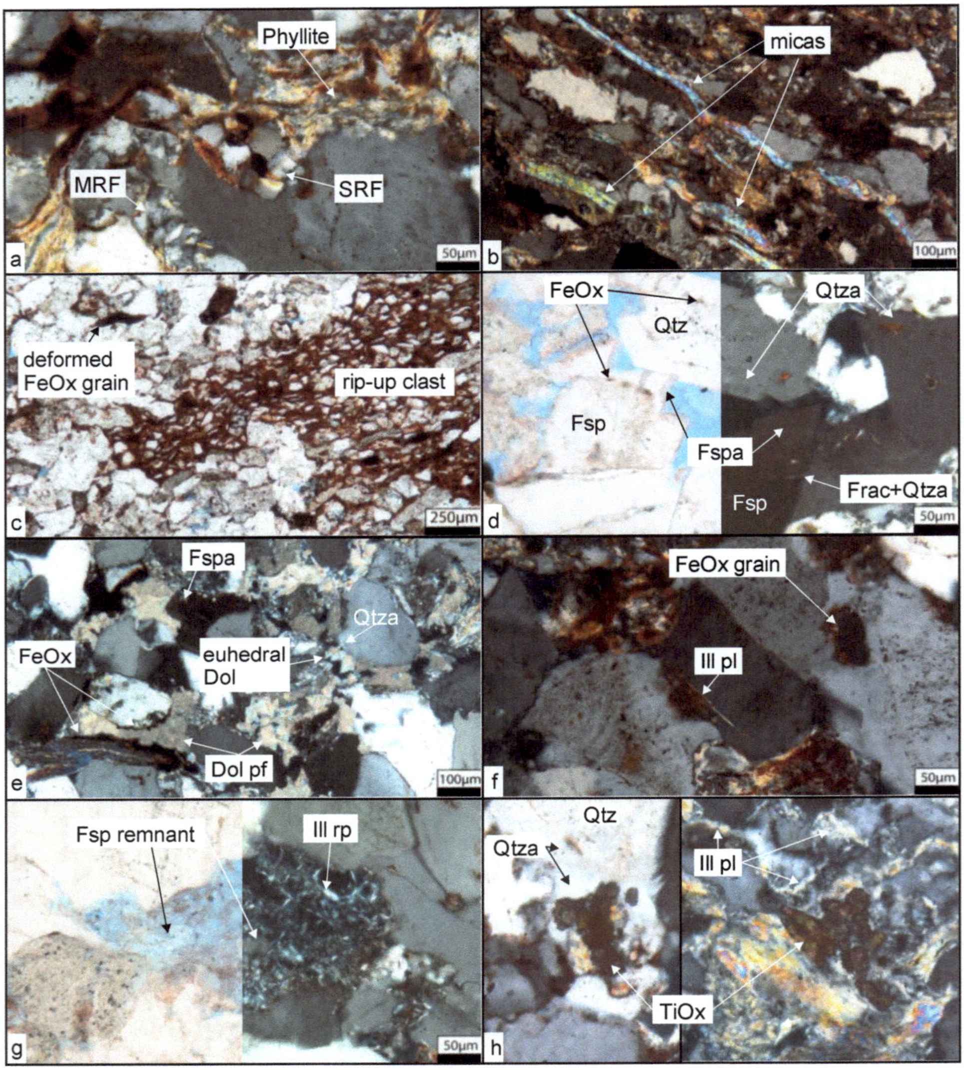

Figure 3.4: Figure caption on the next page.

Figure 3.4: a) Overview of rock fragments, e.g. a sandstone rock fragment (SRF), metamorphic rock fragment (MRF), and a deformed phyllite clast (RB-24, xpl). b) Kinked/deformed detrital mica flakes aligned to the bedding (RB-21, xpl). c) Rip-up clast from the NW-part of the studied outcrop and deformed iron oxide (FeOx) grains (RB-19, ppl). d) Syntaxial quartz overgrowths (Qtza) limited in their growth by euhedral feldspar cement (Fspa), each on the respective substrate grain (Qtz, Fsp) partially enclosing reddish-brown remnants of a hematite dust rim (FeOx). A fracture in a feldspar grain and overgrowth contains quartz cement from the neighboring detrital grain (Frac+Qtza, RB-26, ppl = left, xpl = right). e) Dolomite cement (Dol) sometimes occurs pore-filling directly around detrital quartz and feldspar grains without syntaxial overgrowths, e.g. to the left of the image (Dol pf). However, in the upper half of the image it encloses authigenic quartz and feldspar cements (Qtza, Fspa) and develops euhedral rhombs. FeOx occurs as pore-lining phase on detrital grains (RB-28, xpl). f) Pore-lining illite (Ill pl) on a detrital grain and underneath authigenic cement as well as reddish hematite pigments and iron oxide grains (RB-13, xpl). g) Illite replacing a former feldspar grain with only grain remnants and hematite pigments visible (Ill rp, RB-26, ppl = left, xpl = right). h) Two examples of authigenic titanium oxide (TiOx) forming subhedral crystal aggregates in the intergranular volume. On the left TiOx is encased in quartz cement (RB-25, xpl). On the right the aggregates fill the intergranular volume, which also features abundant pore-lining illite fibers (RB-06, xpl).

Mineral grains include mica, primarily muscovite and some biotite adding up to on average 2.1 % (max. 8.7 %, min. 0.3 %, Fig. 3.4b), and accessory minerals as zircon, tourmaline, and opaque and heavy mineral grains combine to a maximum of 1 % in each sample. Iron oxides are present in all samples and occur as pore-lining and filling phases, grains, and iron-rich pseudomatrix with on average 14 % (max. 21 %, min. 6 %, Fig. 3.4e). Some of the red sandstone samples contain greyish white bleached zones as speckles or bleached laminae.

Reddish-brown rip-up clasts occur as highly deformed hematite-rich shaly siltstone fragments in two thin sections in the NW of the outcrop (RB-17, RB-19, Fig. 3.4c). Detrital matrix was not encountered during point counting.

The iron oxide contents in the SE-part are below the mean, while higher contents are present in the NW-part (Fig. 3.5a). The contents of detrital mica increase towards the NW-part (Fig. 3.5b). The grain sizes show an insignificant deviation from the average grain size of 150 μm along the profiles (Fig. 3.5c). Similarly, correlations are poor for vertical variations in grain size, iron oxide-, and mica contents. The few recognizable sedimentary structures do not appear to correlate with e.g. mica contents (Fig. 3.2a). An above-average mica content in a sample originating from the small channel in the basal coset (red, RB-13) is opposed by a below-average value for the overlying sample within the same channel structure (orange, RB-32, Fig. 3.2a, 5b at x = 14.7 m).

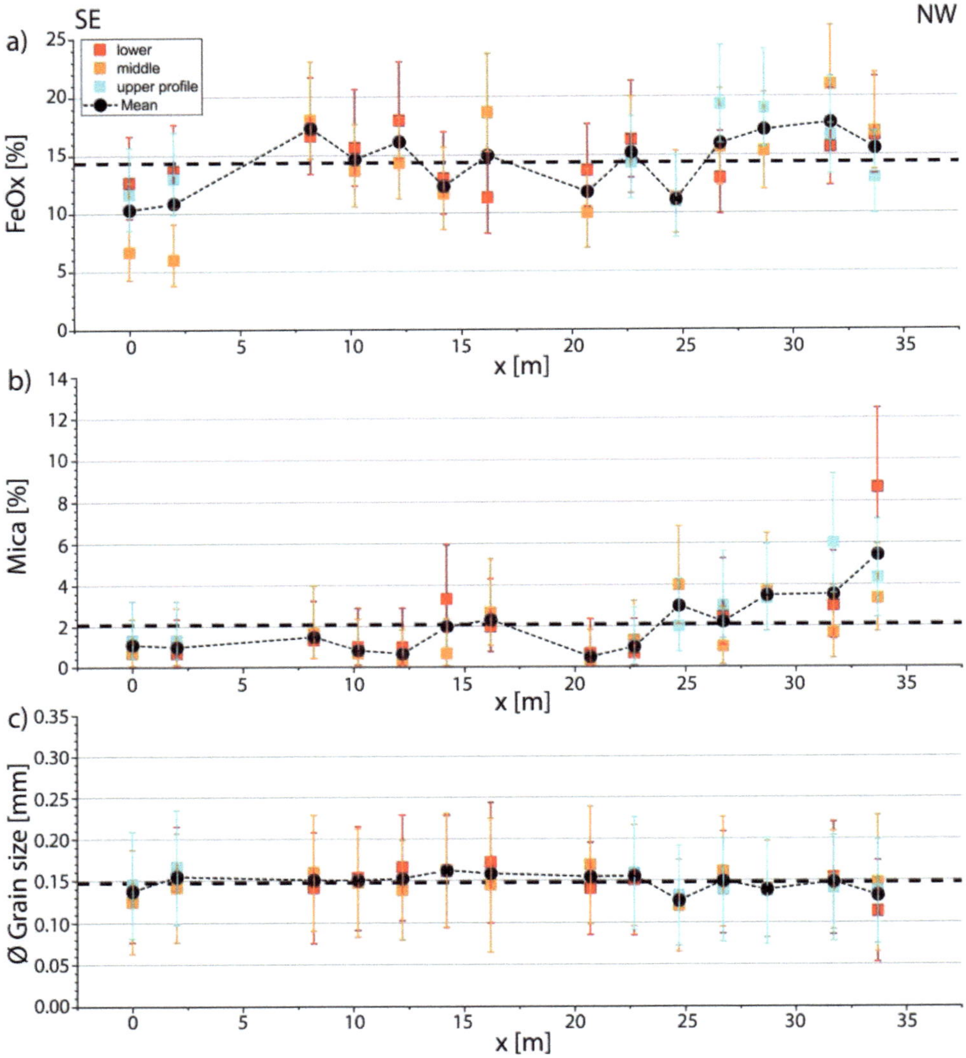

Figure 3.5: Scatter plots along the quarry wall from SE to NW for all profiles (dashed line - average value; dotted line - mean of the three samples at position x). a) Lateral variation of the iron oxide content (FeOx, avg. 14 %, max. 21 %, min. 6 %). b) Lateral variation of the mica content (avg. 2.1 %, max. 8.7 %, min. 0.3 %). c) Lateral variation of the average grain size for all samples (avg. 0.150 mm, max. 0.170 mm, min. 0.110 mm).

Authigenic phases

The most common authigenic mineral in the study area is quartz with on average 15 % (max. 20 %, min. 7 %, Fig. 3.4d, e). Other blocky, pore-filling cements are feldspar with on average 5 % (max. 8 %, min. 2 %, Fig. 3.4d, e) and occasionally dolomite (avg. 0.2 %, max. 1.3 %, min. 0 %, Fig. 3.4e). Feldspars often formed euhedral cements, which occasionally stop neighboring quartz cements and prevent their development into a euhedral overgrowth (Fig. 3.4d). A fracture in a euhedral feldspar overgrowth is filled by quartz cement from a neighboring detrital grain (Fig. 3.4d). Dolomite cements often appear with a pitted crystal outline or are partially dissolved. They form localized pore-filling patches with a diameter of up to several hundreds of µm (Fig. 3.4e). Those completely pore-filling dolomite textures enclose detrital grains without any overgrowth cements and appear to prevent the formation of long grain contacts during compaction (Fig. 3.4e, left). Furthermore, some dolomites show euhedral rhombs, which fill pores and also replace feldspar cements (Fig. 3.4e, top). The dolomite rhombs may occur on euhedral feldspar and quartz cements (Fig. 3.4e, top).

Illite is present both as a pore-lining and a pore-filling phase with on average 7 % (max. 16 %, min. 2 %, Fig. 3.4f, h). The pore-lining phase is present as a grain coating. Minor amounts of kaolinite (avg. 0.03 %) occur as pore-filling booklets in some samples filling intergranular pores, sometimes together with illite fibers. Authigenic titanium oxides (TiOx) are present as clusters of small anhedral aggregates in the pore space (avg. 0.1 %, Fig. 3.4h). Some authigenic minerals also replace existing grains, such as illite replacing feldspars (avg. 1 %, max. 3 %, min. 0 %, Fig. 3.4g) and subordinately kaolinite as well as chlorite replacing micas and rock fragments (avg. 0.2 %, max. 1.0 %, min. 0 %). Kaolinite has been rarely encountered as a replacive phase throughout petrographic analyses but was not registered during point counting.

Lateral variation of quartz cement contents (avg. 15 %, max. 20 %, min. 7 %) along the profiles is small with lower mean values for the three samples in the NW-part (11 to 15 %, Fig. 3.6a, black dots). Illite volumes scatter around the average value of 7 % (max. 16 %, min. 2 %, Fig. 3.6b). The mean values for five sample locations in the SE mostly plot below average (4 to 7 %). Comparably high values for five samples of the NW-part are encountered (8 to 10 %) with one exception of 6 % mean illite content at 31.7 m. Furthermore, the amount of illite volumes exhibits a broader spread in the NW than in the SE along the profile (Fig. 3.6b). This is similar to observations in the cross-plot of mica contents (Fig. 3.5b). The scatter plot displaying feldspar cement contents along the profiles shows no deviation of data from the average (Fig. 3.6c). The indicated correlation between sample position and illite content cannot be linked to sedimentary structures. Four samples from the NW of the outcrop (RB-35, -37, -38, -43), all originating from the same coset with shallowly NW-dipping sets in the topmost profile (blue, Fig. 3.6), plot above and below the average value.

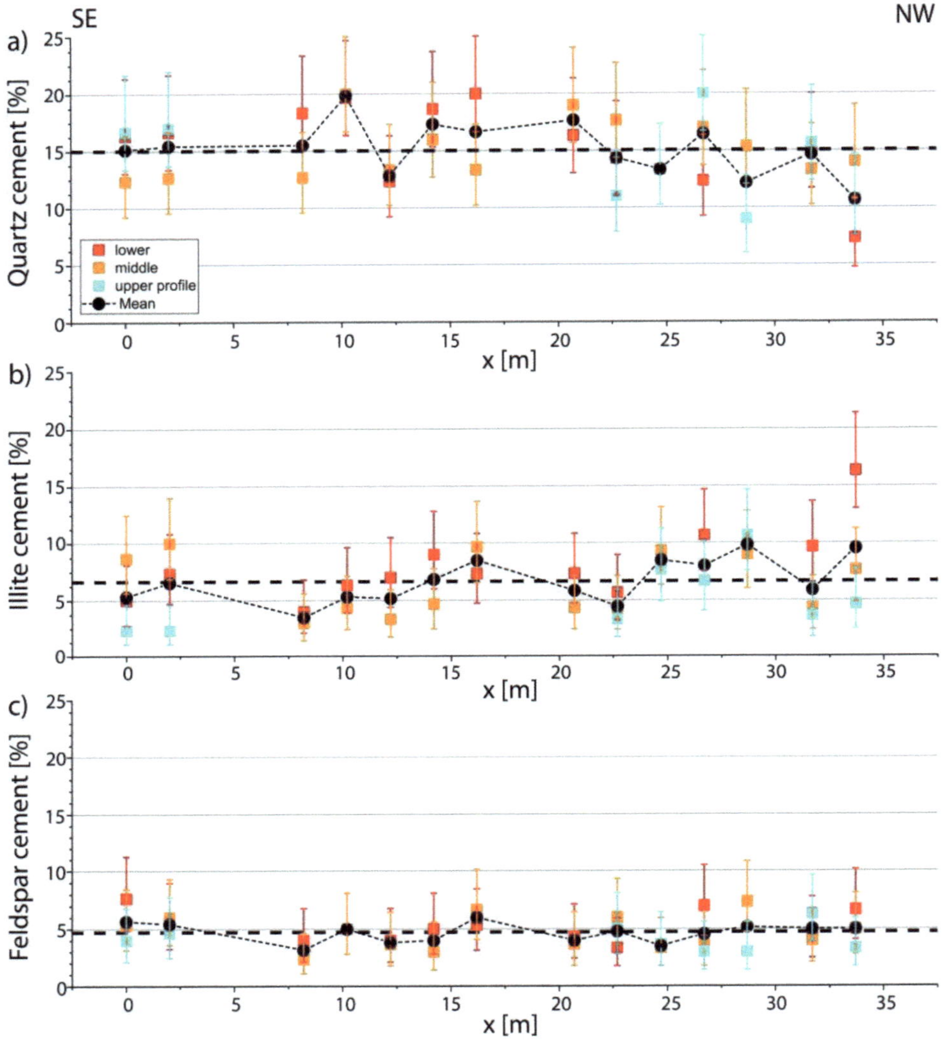

Figure 3.6: Scatter plots along the quarry wall from SE to NW for all profiles (dashed line - average value; dotted line - mean of the three samples at position x). a) Lateral variation of the quartz cement contents (avg. 15 %, max. 20 %, min. 7 %). b) Lateral variation of the illite contents (avg. 7 %, max. 16 %, min. 2 %). c) Lateral variation of the feldspar cement contents (avg. 5 %, max. 8 %, min. 2 %) for all samples.

3.5.3 Spectral gamma ray

Calculated API-values along profile 1 range from 135.8 to 169.2 API (Fig. 3.7). Data generally increase from the SE to the NW. With two exceptions of roughly average gamma ray values (RB-01 at 0 m and RB-05 at 12 m) all samples located between 0 and 22.7 m lie below the average gamma ray value of 150.1 API whereas all following samples from 24.7 m with one exception (RB-18 at 26 m) are characterized by above-average values. The first high spectral gamma ray reading coincides with the plug sample containing rip-up clasts (RB-17, x = 24.7 m). Further towards the NW SGR-values decrease (x = 26.7 m) before attaining continuously high values again from sample RB-19 on, which contains a rip-up clast (x = 28.7 m). Sedimentary structures, e.g. the channel structure in the SW-wall (RB-13, Fig. 3.2a), apparently do not affect the SGR values. None of the extrema of the SGR-plot corresponds to a specific type of sedimentary structure. Absolute values are on average 5.0 % K, 2.7 ppm U, 11.9 ppm Th. In the SE, values are at 4.8 % K, 2.8 ppm U, 12.6 ppm Th and reach up to 5.1 % K, 3.1 ppm U and 15.5 ppm Th in the NW (Tab. A.1).

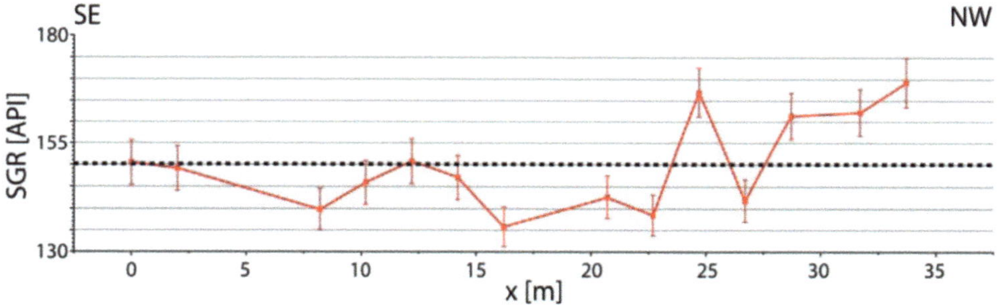

Figure 3.7: Spectral gamma ray-curve of profile 1 (samples RB-08 to RB-21) along the quarry wall. The average spectral gamma radiation is at 150.1 API (dashed line) (max. 169.2 API, min. 135.8 API).

3.5.4 Porosity and permeability

Porosities (Phi) of 36 plugs range from 10.5 to 18.8 % (avg. 15.5 %) (Tab. A.1). Permeabilities (k) range from 0.02 to 48.43 mD (avg. 9.19 mD) (Tab. A.1). The porosity-permeability scatter plot (Fig. 3.8) exhibits a good correlation between porosity and permeability. Overall, both porosity and permeability become smaller and reach 10.5 % and 0.02 mD, respectively, to the NW (Fig. 3.8).

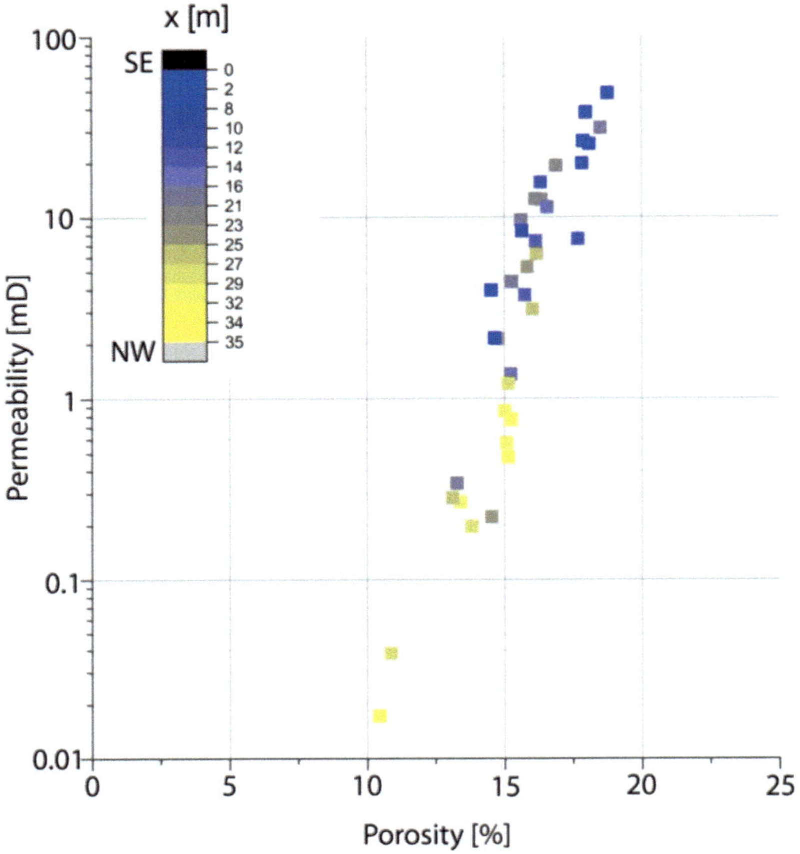

Figure 3.8: Cross-plot of permeability versus porosity with the color code indicating the position of the specific sample along the quarry wall (numbers = distance from SE to NW in meters).

In profiles 2 and 3 porosities are close to the average value of 15.5 % (Fig. 3.9a, b). The largest decrease by more than 5 % is shown in the NW-part of the basal profile 1 (Fig. 3.9c).

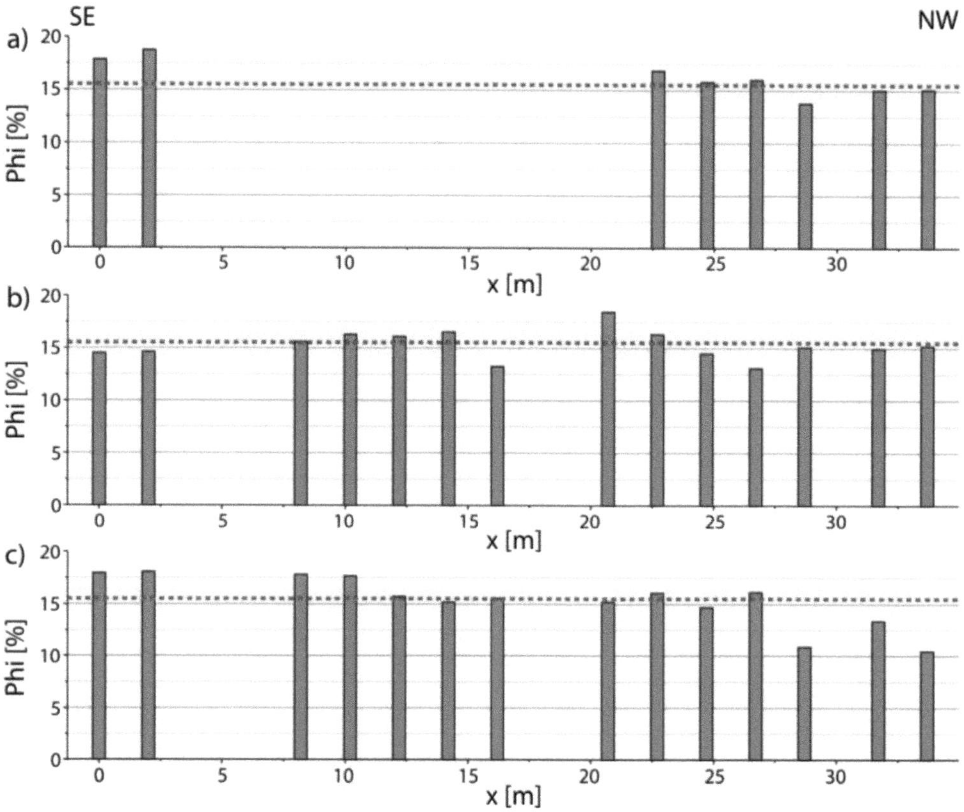

Figure 3.9: Porosity profiles along the quarry wall from SE to NW (a - upper profile 3, b – middle profile 2, c - basal profile 1) with the average porosity of 15.5 % displayed as dashed line (max. 18.8 %, min. 10.5 %). The average standard deviation accounts for 0.2 %.

In profiles 1 and 3, permeabilities vary by about one order of magnitude between 0 to 22.7 m (Fig. 3.10a, c). Starting at 24.7 m permeabilities drop by one order of magnitude with overall k < 1 mD between 28.7 and 33.7 m. Profile 2 exhibits very low permeability at 16.2 m (RB-24, 0.34 mD, Fig. 3.10b) and a sample at 28.7 m that slightly exceeds 1 mD (RB-22, 1.20 mD) but generally match a permeability decrease toward the NW.

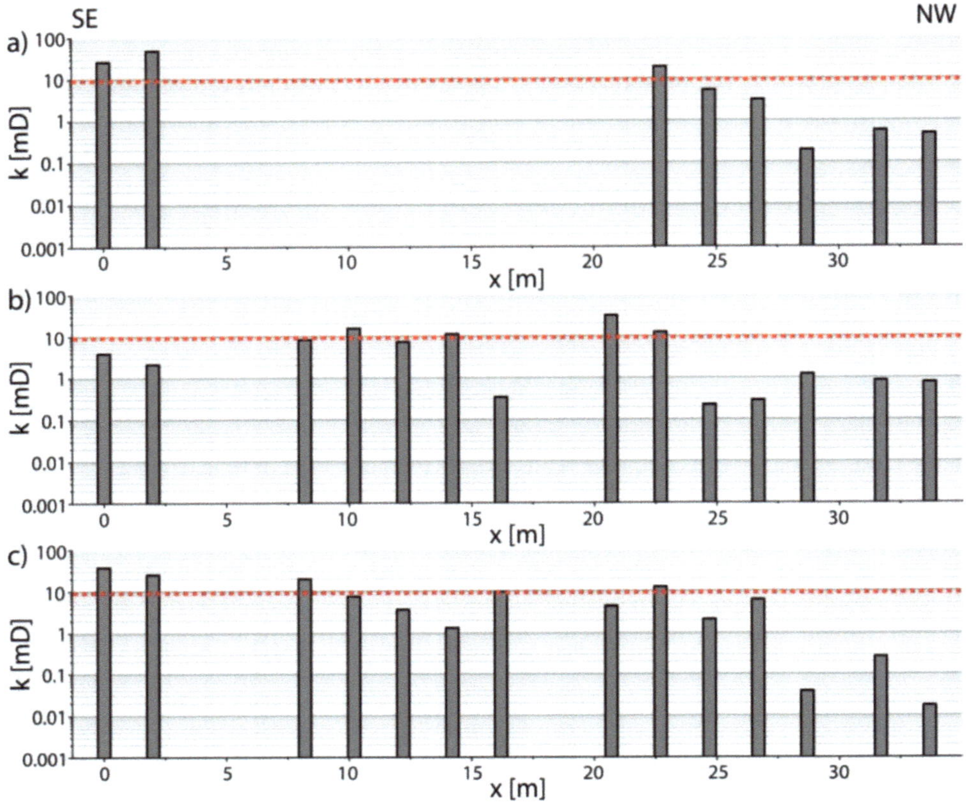

Figure 3.10: Permeability profiles along the quarry wall from SE to NW (a – upper profile 3, b – middle profile 2, c – basal profile 1) with the average value of 9.19 mD displayed as dashed lines (max. 48.43 mD, min. 0.02 mD).

3.5.5 Correlation of reservoir quality controls

To assess the impact of detrital and authigenic constituents on porosity and permeability, scatter plots coded for the content of constituents were created (Fig. 3.11). An increase in bubble size represents an increase in content. Iron oxides correlate poorly with porosity or permeability (Fig. 3.11a). However, with an increased detrital mica content porosity and permeability are reduced (Fig. 3.11b). The cross-plot for authigenic quartz indicates higher porosities and permeabilities with slightly increased cement volumes (Fig. 3.11c). Higher illite contents correlate with lower porosities and permeabilities (Fig. 3.11d).

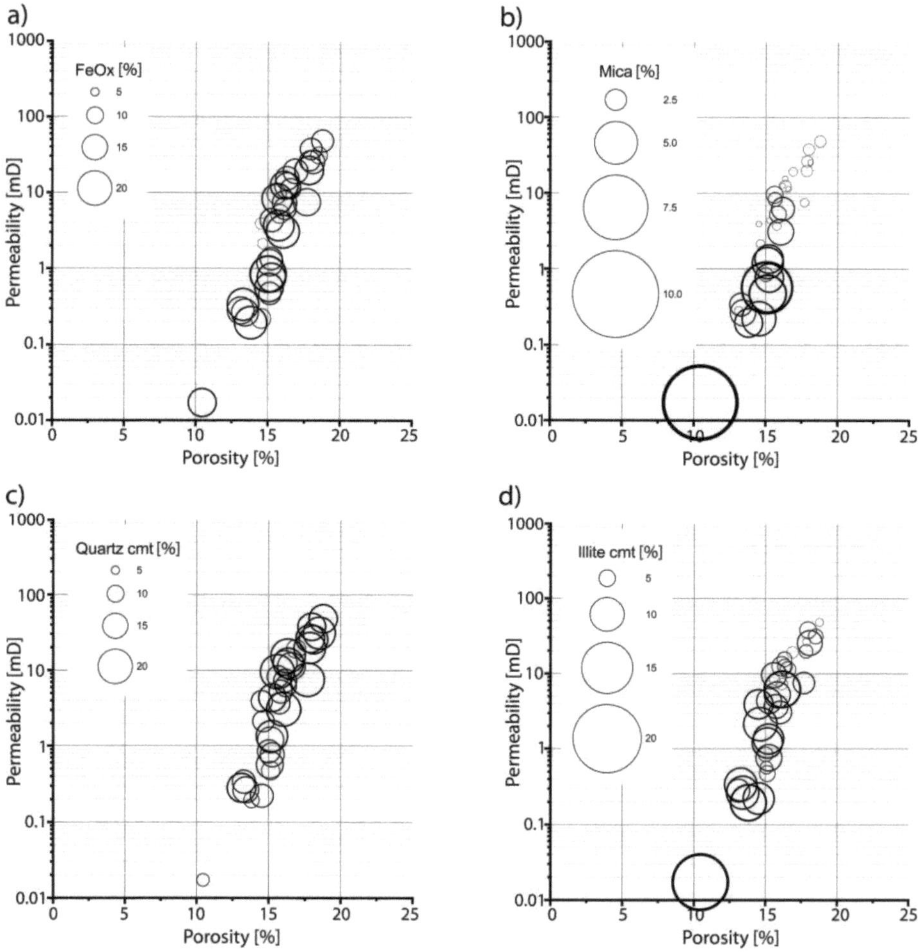

Figure 3.11: Bubble chart exhibiting relationships between porosity, permeability, and mineral phases. a) The contents of FeOx-rich grains poorly correlate with porosity and permeability. b) Samples containing high contents of detrital micas have lower permeability and porosity. c) Quartz cement content shows a poor correlation with porosity and permeability. d) Samples containing higher illite contents have lower porosity and permeability.

3.5.6 Compaction

The average IGV in the studied samples is 33 % (max. 38 %, min. 24 %). Compactional porosity loss (COPL) is on average at 18 % (max. 28 %, min. 11 %) and cementational porosity loss (CEPL) is at 22 % (max. 29 %, min. 13 %). To assess if there is a general correlation of enhanced compaction or cementation deteriorating reservoir quality along the sampled profiles, COPL is plotted against CEPL after Lundegard (1992) (Fig. 3.12).

25 out of 34 samples lose porosity primarily via cementation (Fig. 3.12). The plot shows a poor correlation between the position along the profile from NW to SE and COPL or CEPL (Fig. 3.12).

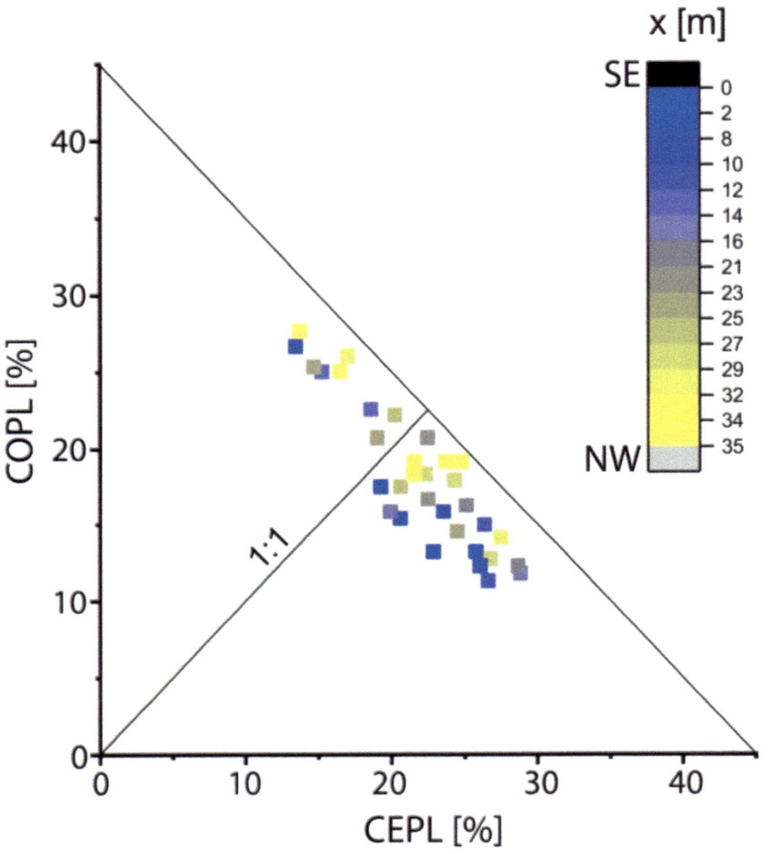

Figure 3.12: Plot after Lundegard (1992) with results plotting both in the cementational porosity loss (CEPL) dominated area (25 out of 36 samples) and compactional porosity loss (COPL) dominated area (11 out of 36 samples) (blue = SE-part of outcrop, yellow = NW-part of outcrop).

Laterally, the plots for IGV, compactional and cementational porosity loss do not show a coherent correlation between COPL-/CEPL-values and the position along the profile from SE to NW (Fig. 3.13). Only the samples from the SE- and NW-sides and their mean for IGV distinctly plot above (+2-3 %, SE) and below average (-3-4 %, NW) (Fig. 3.13a). Similarly, the COPL-values plot below average in the SE (-3 %) and above average in the NW (+3-4 %, Fig. 3.13b).

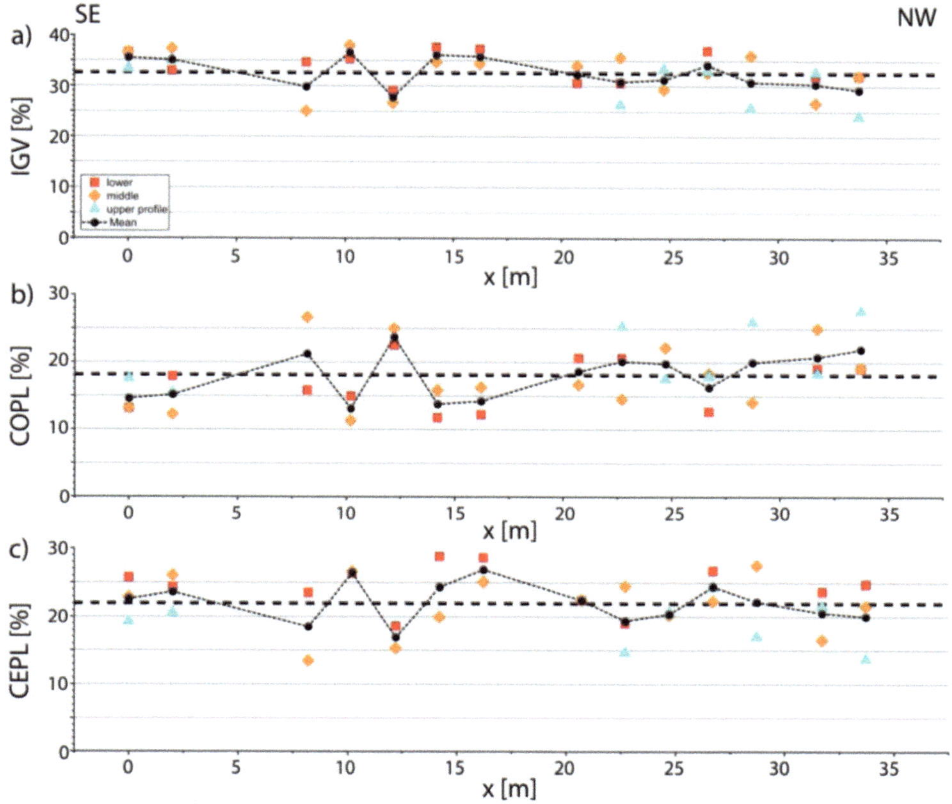

Figure 3.13: Scatter plots along the quarry wall for all profiles (dashed line - average value; dotted line - mean of the three samples at position x). a) The intergranular volume IGV (avg. 33 %, max. 38 %, min. 24 %), b) the porosity loss by compaction (avg. 18 %, max. 28 %, min. 11 %) and c) the porosity loss by cementation (avg. 22 %, max. 29 %, min. 13 %) of all samples along the profiles.

3.6 Discussion

3.6.1 Sedimentology

Very shallow beds gently dipping towards the NW (no. 1, Fig. 3.2a) along the quarry section ending asymptotically at the basal bounding surface in the upper sandstone bed (no. 3, Fig. 3.2a) are interpreted as lateral accretion surfaces (Cadle and Cairncross, 1993; Miall, 1985; Stewart, 1981). They illustrate the ongoing migration of a point bar with the cut bank to the NW following the paleochannel (Bridge, 2006; Heggemann et al., 2003). The erosive character of the basal bounding surface of the upper sandstone bed (no. 3, Fig. 3.2a) is also characteristic for lateral accretion elements (Cadle and Cairncross, 1993; Miall, 1985, 1988b; Stewart, 1981). Its erosive character also prohibits the preservation of a gradational, non-erosive upper bounding surface of the basal sand body, which is another typical aspect of lateral accretion elements but is known to often be truncated by the subsequent sedimentary unit (Cadle and Cairncross, 1993; Miall, 1985). First-order bounding surfaces (no. 1, Fig. 3.2a) outline lenses of a sigmoidal shape typical for lateral accretion features (Marzo et al., 1988). This overall characteristic outline is often obscured by smaller-scale, secondary channels (Miall, 1985) as it is the case in the basal unit featuring two, m-scale channel geometries (no. 2, Fig. 3.2a). Downstream accretion on the m-scale is excluded as mechanism forming those surfaces considering their shallow dip and substantial lateral extent along the 35 m wide quarry section (Ghazi and Mountney, 2009). Furthermore, the Upper Buntsandstein of the region was formed primarily by deposits of meandering river systems (Bock et al., 2005; Freudenberger et al., 2013; Geyer, 2002; Okrusch et al., 2011). Meanders may also exhibit downstream accretion but lateral accretion is more frequently encountered in a laterally migrating sys-tem (Ghazi and Mountney, 2009). Another published example of a meander deposit of the Plattensandstein is the quarry Wüstenzell approximately 10 km East of the study area. Quarry Wüstenzell also features fine-grained, red, cross-bedded sandstone bodies of lenticular shape with few intercalated claystone layers (Bock et al., 2005; Freudenberger, 1990).

The narrow, shallow channels (2 m and 4 m in width) seen in profile in the upper part of the basal bed along the SW-wall (no. 2, Fig. 3.2a) can be assigned to cross-bar channels in the upper point bar deposit of a fluvial meander. Those are mostly discussed to form through discharge fluctuations and shifts in the channel position (Bridge, 2006) or in stages of receding water levels (Sambrook Smith et al., 2016).

The occurrence of subhorizontal bedding at the top of the basal bed along the NW-wall points toward a deposition as either lower or upper plane beds of the corresponding flow regime (Ashley, 1990) (no. 1, Fig. 3.2a). The fine to very fine grain size of on average 150 μm indicates upper plane beds according to bedform stability diagrams (Ashley, 1990; Carling, 1999; Van den Berg and Van Gelder, 1993). The upper plane beds (no. 1, Fig. 3.2a), like the cross-bar channels, are assigned to the top part of a point bar deposit prior to a change in flow velocities or base-level causing the subsequent erosive surface (Bridge, 2006) (no. 3, Fig. 3.2a).

The lower and upper sandstone units are separated by an erosive, subhorizontal surface, which may represent a limited base-level fall or increasing energy in the river system causing successive erosion (no. 3, Fig. 3.2a). Such multi-storey fluvial channels match well with fluvial meanders in a semi-arid environment, as described for the wider region (Bock et al., 2005; Freudenberger et al., 2013; Füchtbauer, 1967, 1974; Mader and Teyssen, 1985; Meschede and Warr, 2019; Okrusch et al., 2011; Paul, 1982; Tietze, 1982).

In the upper bed of the SE-wall, foresets of higher dip angles, having a tangential character and reduced lateral extent, can be interpreted as downstream accretion surfaces (Ghazi and Mountney, 2009) (no. 1, Fig. 3.2a).

The presence of rip-up clasts of silt- and claystones (samples RB-17 and RB-19 in profile 1) near the base of the basal bed indicates that rip-up clasts preferentially occur in the vicinity of the bottomsets of fluvial cosets as also reported in similar settings (Bridge, 2006; Henares et al., 2016a; Komatsubara, 2004; Veiga et al., 2002). Their angular clast morphology suggests an intraformational origin as reworked clasts from a nearby floodplain (Bordy and Catuneanu, 2002; Bridge, 2006; Geyer, 2005; Grecula et al., 2003; Heggemann et al., 2003; Henares et al., 2016a; Komatsubara, 2004; Veiga et al., 2002).

3.6.2 Spectral gamma ray and mineral composition

Overall high SGR values of the Plattensandstein are assigned to a high fraction of fine-grained material and relatively high amounts of micas, illite, and K-feldspar (Tab. A.1). The lateral variation of spectral gamma ray values in the study area indicates differences in the sandstone composition related to those minerals as well as rip-up clast or rock fragments (Fig. 3.7). For the studied quarry walls, the increase of the SGR-signal is well reflected by the increase of detrital mica contents towards the NW in petrographic samples (Fig. 3.5b). Micas are reported to be abundant in fluvial deposits and cause higher SGR values (North and Boering, 1999). Their content has been proven to occasionally be higher in floodplain deposits when compared to channel sandstones, thus SGR-signals may be characteristic for floodplain and channel. Signals may also overlap for those two facies obscuring an increase in mica content toward the floodplain (North and Boering, 1999). A decrease in flow velocities favoring the deposition of sheet silicates within the channel may explain the increase of detrital mica to the NW. This hypothesis is supported by the sedimentological structures: Surfaces interpreted as lateral accretion (no. 1, Fig. 3.2a) illustrate a progressive channel migration with assumed increasing curvature of the meandering channel prior to its eventual abandonment. Increasing curvature coincides with reduced current energy of the river (Blanckaert and de Vriend, 2010; Chang, 1983) and thus possibly increased mica deposition. As the overall sedimentary texture does not change, it can be assumed that such velocity reduction is small.

Centimeter-scale rip-up clasts presumably originating from a nearby floodplain occur more frequently in the NW-part of the studied quarry. Thus, an increased gamma ray signal related to detrital mica content and including the presence of clayey to silty rip-up clasts might indicate an increase in eroded intraformational sediments.

3.6.3 Paragenetic sequence

Based on the textural observations for authigenic cements (Fig. 3.4d-h) and comparison with published data, a relative paragenetic sequence can be established (Fig. 3.14).

The earliest diagenetic alteration is interpreted to be the infiltration of grain coating clay minerals and sediment reddening by pigmented hematite rims (Beyer et al., 2014; Busch et al., 2017; Morad et al., 2010; Walker, 1967a, b, 1975; Walker et al., 1978). They frequently occur at grain contacts and are overgrown by all other authigenic phases (Fig. 3.4d, f, g).

An early diagenetic feldspar (and rock fragment) dissolution enables the subsequent formation of feldspar cements. Those euhedral feldspar cements are interpreted to grow during early diagenesis in arid depositional environments (Beyer, 2015; Beyer et al., 2014; Füchtbauer, 1974; Gaupp and Okkerman, 2011; Morad et al., 2010) and are partially encased by subhedral quartz overgrowths (Fig. 3.4d). Consequently, feldspar cements are interpreted to predate quartz cementation. The observation of quartz cement filling a fracture in a euhedral feldspar overgrowth further supports this interpretation, which is in accordance with results of other authors for the Buntsandstein in Central Germany (Füchtbauer, 1967, 1974).

For dolomite an early diagenetic formation is described in literature (Worden et al., 2018; Worden and Burley, 2003). Based on textural relationships and the replacement of feldspar cements in the presented samples, this can be interpreted for parts of the studied dolomite cements. The pore-filling cement texture, which appears to stabilize the grain framework, prevents long grain contacts and encloses detrital grains (Fig. 3.4e, left). From this observation an early diagenetic origin is interpreted. It is supported by the fact that no early or burial diagenetic, syntaxial cement phase like feldspar and quartz are observed underneath the pore-filling dolomite. A burial diagenetic or uplift-related formation of dolomite rhombs (Fig. 3.4e, top) is interpreted where quartz and feldspar cements are enclosed or replaced by those rhombs, thus postdating both cement phases. The pitted appearance of both dolomite cement textures is interpreted to be a result of uplift-related dissolution by meteoric waters (Fig. 3.4e) (Al-Ramadan et al., 2005; Bjørlykke, 2014). Staining did not reveal compositional differences between the observed dolomite cements.

The dissolution of K-feldspar grains and their replacement by illite and alteration to kaolinite could be assigned to different processes (Fig. 3.4g) (Weber and Lepper, 2002). The formation of kaolinite is favored during early diagenetic weathering of feldspar grains when in contact with meteoric waters (Becker et al., 2017; Bjørlykke, 2014; Füchtbauer, 1974; Lanson et al., 2002; Morad et al., 2010; Worden and Morad, 2003). The stability of kaolinite is largely related to kinetic constraints and the K^+/H^+ activity ratio during diagenesis (Lander and Bonnell, 2010; Lanson et al., 2002). Pore-filling kaolinite occurrence is low in the presented samples, which can be assigned to pervasive subsequent illitization with available K^+ from feldspar dissolution or from external sources at low acidity (Busch et al., 2020; Lanson et al., 2002). Thus, a sufficient supply of K^+ is implied as well as high precipitation rates at 120-140°C with precipitation starting already at lower temperatures during burial (Bjørlykke, 1998). Additionally, arid conditions at early burial may have also slowed kaolinite formation during early diagenesis, as observed in fluvial sandstones in an arid environment during the Upper Carboniferous (Becker et al., 2019; Becker et al., 2017). The replacement of feldspars by illite occurs prominently along the twin planes (Busch et al., 2019; Molenaar et al., 2015). Illitization of feldspars is interpreted as a burial diagenetic alteration.

The illitization of infiltrated, likely smectitic and hematite-stained, clay mineral grain coatings is interpreted to follow the early diagenetic phase of K-feldspar dissolution and kaolinite formation, supplying the K^+-ions. The illitization also likely predates quartz cementation since illitized coatings are overgrown by syntaxial quartz cements (Fig. 3.4d). A burial diagenetic replacement of smectitic precursor clays is also reported for other sandstones deposited in arid depositional systems (Liewig and Clauer, 2000). The replacing reactions of kaolinite and illite after feldspar and kaolinite, respectively, produce minor amounts of excess silica, which could source the formation of quartz cements (Beyer et al., 2014; Morad et al., 2010; Weber and Lepper, 2002). Illitized parts of feldspar grains remain as a meshwork within intragranular pores (Fig. 3.4g).

Quartz cementation likely syn- to postdates mechanical and chemical compaction, since samples containing a compacted fabric contain visually less quartz cements (Fig. 3.4a, b) than less compacted samples (Fig. 3.4d). Sources for quartz cements could be intraformational pressure dissolution resulting in serrated grain contacts (Fig. 3.4b) (Walderhaug and Bjorkum, 2003), clay mineral transformations or feldspar replacements and dissolution (Lander and Walderhaug, 1999; Morad et al., 2010; Worden and Morad, 2003). Quartz cementation is generally described to be a burial diagenetic alteration, being pervasive at temperatures above 70-75°C (Lander and Walderhaug, 1999; Morad et al., 2010; Taylor et al., 2010; Walderhaug, 1996, 2000) and kinetically controlled if the pore fluid is near the saturation with SiO2 (Rimstidt and Barnes, 1980; Walderhaug, 1996).

Quartz cements are absent in some intragranular pores in feldspars that were partially replaced by illite and maintaining the original grain outlines (Fig. 3.4g). This dissolution is interpreted to postdate syntaxial quartz cementation and illitization. It likely coincides with uplift diagenetic dissolution of unstable detrital grains and partial dissolution of dolomite cements when flushed by meteoric fluids (Becker et al., 2017; Wüstefeld et al., 2017) (Fig. 3.4e).

The formation of minor amounts of TiOx cements is often related to the dissolution and alteration of Ti-bearing detrital mineral grains (e.g. ilmenite, sphene, or biotite) (Füchtbauer, 1974; Morad and Aldahan, 1986; Sindern et al., 2019). Anatase dating in Permian Rotliegend sandstones from the NW-German Basin, deposited in an arid depositional environment, by Sindern et al. (2019) support a timing prior to illite precipitation but state that both illite and TiOx formed during the initial stages of burial diagenesis. Generally, the diagenetic formation of pore-filling titanium oxide aggregates is known to occur as an early to burial diagenetic phase (Morad and Aldahan, 1986; Morad and Aldahan, 1987). As the few titanium oxide aggregates in the samples do not appear strongly affected by compaction a burial diagenetic formation is interpreted. As aggregates occur surrounded by quartz cement, they are interpreted to predate or coincide with the formation of authigenic quartz supporting an early burial diagenetic formation (Fig. 3.4h, 3.14).

Minor amounts of chlorite replacements of rock fragments and micas do not show a distinct paragenesis with other minerals but are reported to be prominent burial diagenetic replacements of ferromagnesian-rich grains (as e.g. volcanic rock fragments or biotite) (Anjos et al., 1999; Worden and Morad, 2003).

The bleaching of thin laminae within the sandstones, discoloring the originally red sandstone to a whitish color, is interpreted to be due to the alteration of grain rimming iron oxide phases (Beitler et al., 2003; Chan et al., 2000; Hilse et al., 2010). The alteration is interpreted to postdate the first dolomite cementation preserving substantial iron oxide rims (Fig. 3.4e). The timing of bleaching is interpreted to coincide with early diagenetic feldspar cementation and the initial stages of quartz cementation since bleached areas do still exhibit minor iron oxide pigments on grain outlines overgrown by syntaxial, blocky feldspar- and quartz cements (Fig. 3.4d) (Schmidt et al., 2020a).

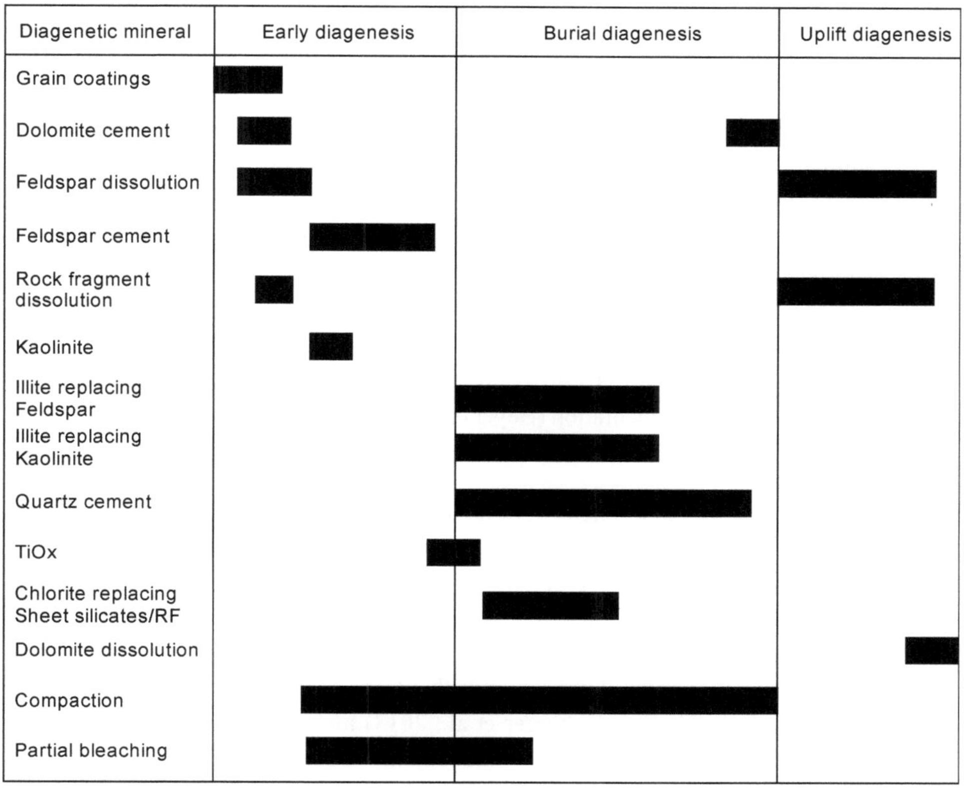

Figure 3.14: Paragenetic sequence of the study area.

3.6.4 Reservoir quality controls

Sandstone permeability varies over almost four orders of magnitude (0.02 to 48.43 mD) over distances of 33.7 m laterally and 1.2 m vertically in this study (Fig. 3.10, 3.12). Permeability variations in other fluvial deposits are reported to also extend over three orders of magnitude due to sedimentary structures, lithological variability, and compaction of ductile rock fragments (Becker et al., 2017; Hartkamp-Bakker and Donselaar, 2009; McKinley et al., 2011; Wachutka and Aigner, 2001). The observed permeability decrease toward the NW-part of the profiles can be related to compositional differences (Fig. 3.5, 3.6, 3.11). While rocks contain slightly higher mean FeOx grain contents to the NW (Fig. 3.5a), Figure 3.11a illustrates that the amount of FeOx grains is not linked to porosity or permeability. For feldspar cements the scatter plot (Fig. 3.6c) displays that cement quantity is not directly related to the measured porosity or permeability. The mean mica content of all profiles increases from 24.7 m onwards as the permeabilities show below-average values (Fig. 3.5b, 3.10). The detrital mica content is negatively correlated with porosity and permeability (Fig. 3.11b). This can be explained by micas being prone to mechanical compaction and enhancing chemical compaction (pressure dissolution) of detrital quartz grains (Becker et al., 2017; Bjørkum et al., 1993; El-ghali et al., 2006; Greene et al., 2009; Henares et al., 2016b; Kristiansen et al., 2011; Morad et al., 2010). Similarly, increased contents of authigenic illite filling the pore space promote pressure dissolution of quartz at grain contacts (Fig. 3.4h). This can be related to reduced permeability (Fig. 3.6b, 3.10, 3.11d) towards the NW-part, which is in accordance to other outcrop analog and subsurface studies (Bjørlykke, 2014; Gaupp et al., 1993; Monsees et al., 2020; Morad et al., 2010; Worden and Morad, 2003). No clear correlation can be established for quartz cements and reservoir quality (Fig. 3.11c). Samples in the NW with highest mica (Fig. 3.5b) and illite contents (Fig. 3.6b) exhibit larger COPL-values (Fig. 3.13b) related to enhanced chemical compaction prior to quartz cementation (Beyer et al., 2014; Greene et al., 2009; Kristiansen et al., 2011; Monsees et al., 2020).

Mica contents (Fig. 3.5b) increasing toward the NW and the occurrence of rip-up clasts in the NW may suggest more intraformational erosion and re-deposition of such clasts and dispersed clayey material from a floodplain (Fig. 3.15). This results in increasing API-values towards the NW (Fig. 3.7) and is reflected by an increasing mean of COPL-values due to chemical compaction (Fig. 3.13b) (Henares et al., 2016b). From sedimentary structures it is interpreted that channel migration in this meandering system (e.g. Bock et al., 2005; Okrusch et al., 2011; Freudenberger et al., 2013) took place toward the NW (Fig. 3.2, 3.15). The paleochannel centers of the basal and the upper sandstone unit are expected to lie further to the NW prior to abandonment (Fig. 3.15). Thus, the illite contents increasing to the NW may originate from increased infiltration of precursory, detrital clay near the floodplain or from the proximity of an overlying, abandoned paleochannel (Moraes and Surdam, 1993). The deposition of micas (Fig. 3.5b, profile 3) and clay particles, which during burial diagenesis undergo illitization, to the NW can be related to an overall decreasing flow velocity of the river system with lateral accretion and increasing channel curvature.

The outcrop is representative for the Plattensandstein of the region and results may be applicable to explain reservoir quality variations on a regional scale. Similar sampling at equivalent outcrops in the wider study area could support the findings of this study.

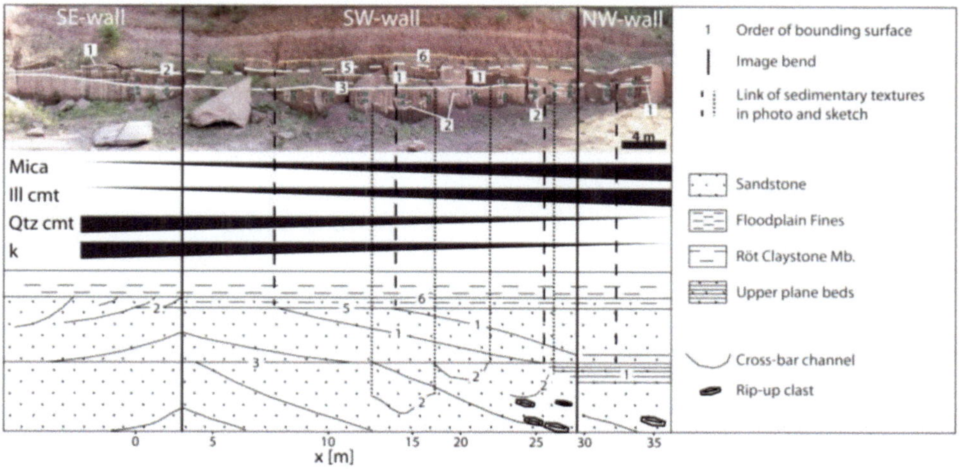

Figure 3.15: Composite image showing sedimentary structures in the panoramic photo and the sedimento-logical sketch. Black triangles display the lateral distribution of rock properties influencing reservoir quality (triangular bars indicating increases and decreases in mica, illite, and quartz content as well as permeability k).

3.7 Conclusion

The outcrop exposes a Lower Triassic multi-storey fluvial meander with point bar deposits and cross-bar channels. The two fine-grained sandstone beds are separated by an erosive surface and feature lateral accretion surfaces.

The variation in porosity (10.5 to 18.8 %, avg. 15.5 %) ranges over 8 % and permeability (0.02 to 48.43 mD, avg. 9.19 mD) ranges over three orders of magnitude over 35 m laterally and 1.2 m vertically in the fine-grained sandstones.

The lateral decrease of permeabilities to the NW is reflected in all three profiles and can be related to an increased content of detrital mica and authigenic illite. Furthermore, the lateral variation of mica and illite is well reflected in spectral gamma ray signatures with an increase of 19 API from the average value.

The increased amount of detrital mica to the NW of the study area enhances chemical compaction and reduces available intergranular pore space and permeability. The amount of illite may add to chemical compaction and porosity reduction.

4 Lithofacies control on compaction and reservoir quality in Lower Triassic fluvial red beds, S-Germany

4.1 Abstract

Reservoir quality is often related to depositional rock types and lithofacies. In this study two lithofacies types within a Lower Triassic outcrop analog of the fluvial Vogesensandstein Formation at the eastern flank of the Upper Rhine Graben are analyzed. The corresponding Buntsandstein Group is a known reservoir lithology in the Upper Rhine Graben and Southern Germany, for geothermal energy, hydrocarbons, and potentially for storage. In the quarry in Lahr-Kuhbach 30 m high walls crop out featuring two lithofacies types of which 1.2 m of cross-bedded, medium-grained sandstones (St) and 0.8 m of fine-grained wavy- to ripple-bedded units (Swr) are the focus of this study to delineate reservoir quality controls. Both lithofacies types are analyzed with regard to sandstone structure, texture, porosity, and permeability as well as mineralogy and its diagenetic evolution.

Lithofacies type St exhibits a slightly larger average grain size than Swr samples (by 0.05 mm) coinciding with lower spectral gamma ray measurements (by 19.9 API) and lower grain coating coverages (by 27 %). All of those parameters indicate slightly lower flow velocities of the river during Swr deposition leading to a higher input of detrital clays into the sediment. Smaller grain coating coverages and higher quartz cement volumes are found in the St samples and stabilize their grain framework as opposed to Swr samples. Grain contact types, intergranular volumes, and compactional porosity loss indicate a higher degree of chemical compaction within the Swr samples that correlates well with the higher grain coating coverages and smaller syntaxial overgrowth cement volumes. Compaction is the main control on reservoir quality leading to Swr samples exhibiting a lower average permeability by four orders of magnitude (0.09 mD vs. 236 mD).

Characterizing lithofacies types can thus offer a means to classify rock units according to differing styles of sedimentation and subsequent diagenetic overprint.

4.2 Introduction

As a consequence of structure in reservoir sandstones, it is assumed that petrophysical properties need to be determined in three perpendicular directions to appropriately mirror subsurface conditions (Meyer and Krause, 2006; Pranter et al., 2007). The heterogeneity detected in sandstones is difficult to grasp from subsurface data as e.g. well spacing often exceeds the lateral dimensions of architectural elements within a fluvial system hosting specific sandstone structures (Pranter et al., 2007). Permeability measured perpendicular to horizontally laminated or cross-bedded fluvial deposits is generally considered to be lower than parallel to lamination or bedding, due to the common gradation of sediments (Allen, 1965; Becker et al., 2019; Becker et al., 2017; Bridge, 2006; Ghazi and Mountney, 2009; Pranter et al., 2007). Furthermore, less permeable sedimentary features may act as vertical baffles, as fluvial deposits commonly contain channel lags comprising mud clasts (Bridge, 2006; Doyle and Sweet, 1995; Henares et al., 2016a), mud drapes on inclined strata, fine-grained ripple intercalations or floodplain deposits (Bridge, 2006; Doyle and Sweet, 1995; Hartkamp-Bakker and Donselaar, 2009; Hornung and Aigner, 1999; Pranter et al., 2007). Other studies propose that permeability perpendicular to bedding or lamination may constitute a more homogenous and efficient fluid displacement front while producing from a reservoir (Meyer and Krause, 2006). This is due to a homogenous baffling when crossing low permeability-laminae compared to a heterogeneous displacement resulting from fluid flow along laminae of differing permeabilities (Meyer and Krause, 2006).

Reservoir qualities (porosity and permeability) have been linked to lithofacies associations of braided or meandering river deposits. The latter exhibit more finely laminated (Sl) or rippled sedimentary structures (Sr) than cross-beds (St) and lower permeabilities but similar porosities (Wachutka and Aigner, 2001). Other studies show higher permeabilities in sandy lithofacies (Sm, Sh, Srt, St after Miall, 1996) than in muddy or silty lithofacies (Fl, Fms after Miall, 1996) (Hornung and Aigner, 2004). Sandy lithofacies Sr represents the exception with relatively low permeabilities (Hornung and Aigner, 2004). That observation was proven in a different stratigraphic unit by Hornung and Aigner (1999) and included the lithofacies type Sl exhibiting lower peak permeability.

An alternative to a categorization into lithofacies types is the mere distinction into reservoir and non-reservoir units and their association with specific mineralogical properties that in turn reveal compaction and cementation mechanisms influencing the pore space (Lundegard, 1992; Paxton et al., 2002). Such properties may be detrital ductile clasts, micas or clay minerals, which are relevant for mechanical as well as chemical compaction deteriorating reservoir quality (Becker et al., 2017; Bertier et al., 2008; Beyer et al., 2014; Bjørkum et al., 1993; Busch et al., 2017; Fischer et al., 2012; Gaupp et al., 1993; Henares et al., 2016a; Molenaar et al., 2015; Morad et al., 2010; Walderhaug, 1996, 2000). Chemical compaction (pressure dissolution) reduces the available intergranular volume (Beyer et al., 2014; Bjørkum et al., 1993; Bjørlykke, 2014; Gaupp et al., 1993; Lundegard, 1992; Monsees et al., 2020; Morad et al., 2010; Paxton et al., 2002). Grain coating illite has been shown to enhance chemical compaction between quartz grains and thus compaction (Monsees et al., 2020). At the same time illitic grain coatings can inhibit the syntaxial mineral precipitation of authigenic phases on respective substrate grains (Busch et al., 2017; Heald and Larese, 1974; Taylor et al., 2010) potentially lowering cementational porosity loss. Infiltrated grain coatings are present in recent and ancient fluvio-eolian sediments deposited in semi-arid climates (Busch et al., 2020; Molenaar and Felder, 2018). Early diagenetic cement phases such as feldspar and carbonate cements can stabilize the grain framework during mechanical compaction. Where those cements are dissolved at depth or during uplift, reservoir quality may be increased (Beyer et al., 2014; Canham et al., 1996; Henares et al., 2016b; Line et al., 2018; Lü et al., 2015; Paxton et al., 2002; Scherer, 1987; Schmidt et al., 2020a, b; Walderhaug, 1996, 2000). Coevally, pore-filling blocky cements like feldspar, carbonate, and quartz cements may reduce reservoir quality (Bjørlykke, 2014; Busch et al., 2019; Gaupp et al., 1993; Morad et al., 2010; Taylor et al., 2010; Walderhaug, 1996, 2000). Clay minerals like kaolinite may reduce porosity, while illite precipitation instead is linked to reduced permeability (Bjørkum et al., 1993; Gaupp et al., 1993; Heald and Larese, 1974; Morad et al., 2010; Taylor et al., 2010).

The assumed relationship between lithofacies type and petrophysical properties so far has not been investigated regarding detrital and authigenic mineralogy as well as processes affecting the intergranular pore space on the microscopic scale. We combine lithofacies types with the overall mineral assemblage and show that permeability can be controlled primarily by lithofacies, thus depositional processes, and subsequent diagenesis rather than by e.g. relative directions to bedding.

4.3 Geological setting

The study area is an active quarry situated 200 m North of Lahr-Kuhbach East of the Upper Rhine Graben in Southern Germany at the southern flank of the mountain Altvater (410 m a.s.l.) (Regierungspräsidium Freiburg - Landesamt für Geologie Rohstoffe und Bergbau, 2019). It is part of a tilted block, which resulted from the Tertiary graben formation and forms the western margin of the Middle Black Forest (Günther, 2010). While the Black Forest consists mainly of Carboniferous granites and Precambrian-Silurian gneisses representing the crystalline basement, the margin contains Permian Rotliegend and Zechstein as well as Triassic Buntsandstein and Muschelkalk sedimentary rocks (Fig. 4.1) (Günther, 2010; Keßler and Leiber, 1994). Those deposits belong to the marginal facies SW of the epicontinental Germanic Basin (Günther, 2010; Hagdorn et al., 2009; Röhling and Lepper,

2013; Ziegler, 1990). They developed on top of the Variscan basement, which was levelled out due to an extensional regime acting on the local swells while local troughs were coevally filled by Rotliegend sediments at the end of the Permian (Günther, 2010). Successive subsidence of those different minor swells and troughs in the marginal Germanic Basin allowed for increasing but deviating thicknesses of the sands and pebbles that were deposited on top during the Buntsandstein (Günther, 2010; Hagdorn et al., 2009; Röhling and Lepper, 2013; Ziegler, 1982). They were sourced from the Ardennes and Gallic Massif to the W as well as the Vindelician High to the SSE (Backhaus, 1974; Günther, 2010; Hagdorn et al., 2009; Röhling and Lepper, 2013; Ziegler, 1982).

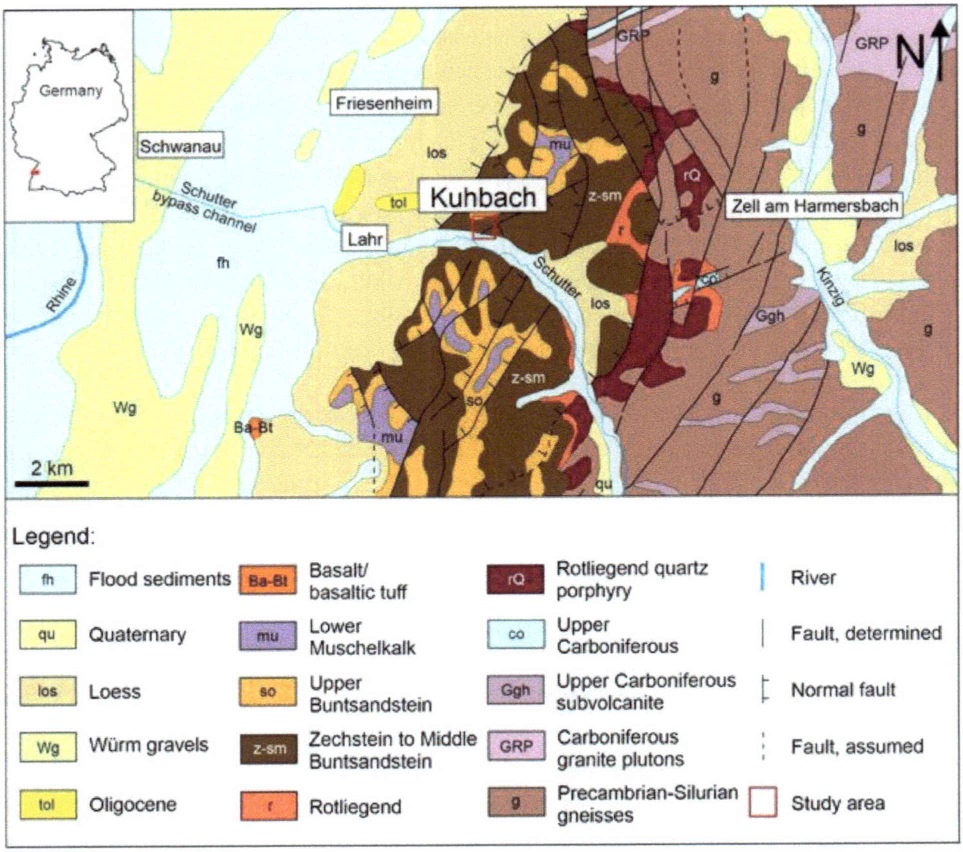

Figure 4.1: Location of the study area on the geological map of Lahr, situated on the eastern shoulder of the Upper Rhine Graben, redrawn from Keßler and Leiber (1994) and Regierungspräsidium Freiburg - Landesamt für Geologie und Rohstoffe (2019).

The Buntsandstein is characterized by a desert-like, semi-arid climate exhibiting major river systems shedding continental alluvial and fluvial deposits toward the North (Backhaus, 1974; Günther, 2010; Hagdorn et al., 2009; Röhling and Lepper, 2013; Ziegler, 1990). Intermittent flooding events occurred, preserved as mud cracks or wave ripples in the clay-rich deposits (Günther, 2010; Ziegler, 1982). The marginal facies exhibits sandy conglomerates to sandstones for the Lower Buntsandstein, whose sedimentary structures like mud cracks and current marks report fluvial channel and limnic deposition (Eck Formation, Fig. 4.2) (Günther, 2010; Hagdorn et al., 2009). It is overlain by the Vogesensandstein Formation, an occurrence exclusive to the marginal facies and covering the Lower and the Middle Buntsandstein (Fig. 4.2) (Günther, 2010; Leiber and Bock, 2013). It contains the Badischer Bausandstein Member, which is present in the studied outcrop (Werner et al., 2013). In the Early and Middle Buntsandstein (Induan to early Olenekian) the marginal facies was characterized by braided rivers (Backhaus, 1974; Paul, 1982; Werner et al., 2013). The fluvial system changed in the Upper Buntsandstein toward primarily meandering streams that exhibited laterally more extensive point bar deposits as well as silty to clayey floodplains (Werner et al., 2013). The Vogesensandstein Formation may locally feature two violet paleosols (VH1 and 2) from episodic pedogenesis (Hagdorn et al., 2009; Keßler and Leiber, 1994; Leiber and Bock, 2013; Werner et al., 2013). VH2 marks the transition to the Upper Buntsandstein beginning with the Plattensandstein Formation of limnic and fluvial origin for the marginal facies (Fig. 4.2) (Günther, 2010; Leiber and Bock, 2013). It contains more paleosol horizons and evolves into the Rötton Formation that represents the onset of marine incursions from the Tethys via the Silesian and East Carpathian Gateways (Aigner and Bachmann, 1992; Schröder, 1982; Ziegler, 1982) followed by the formation of the Muschelkalk Sea (Günther, 2010; Hagdorn et al., 2009; Röhling and Lepper, 2013).

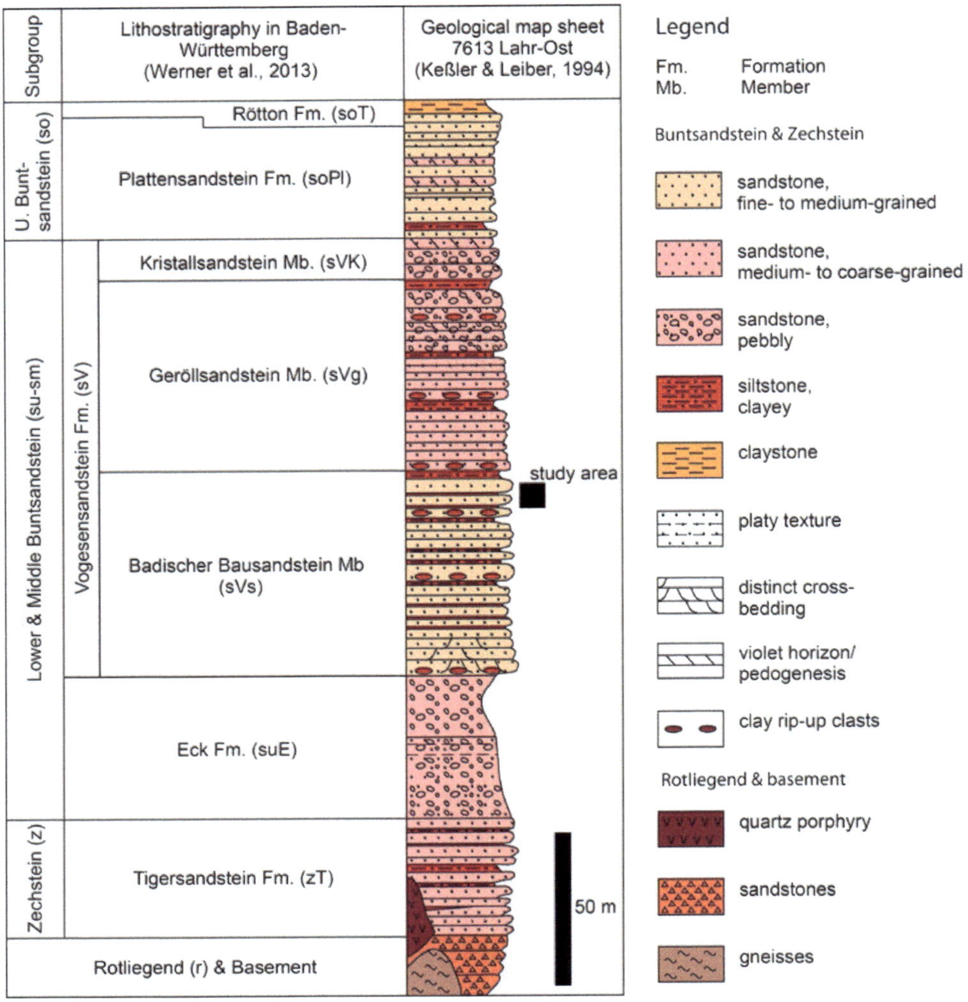

Figure 4.2: Lithological column of the study area (modified from Keßler and Leiber, 1994, and lithostratigraphic classification after Werner et al., 2013). The samples of the studied profiles are derived from the top of the Badischer Bausandstein Member of the Vogesensandstein Formation (black square).

Barite veins are known to occur throughout the Black Forest as well as fluorite or quartz veins (Brockamp et al., 1994; Dill, 1988; Günther, 2010; Hofmann and Baumann, 1984; Staude et al., 2009; Staude et al., 2011; Werner and Dennert, 2004; Werner et al., 2002). Among them, the Diersburg-Reichenbach lode consisting of quartz, barite, hematite etc. is situated less than 5 km East of the study area striking approximately NNE-SSW (Keßler and Leiber, 1994). Veins may date back to the late Variscan orogeny (Dill, 1988; Staude et al., 2011), to the Jurassic to Cretaceous related to the Atlantic opening (Dill, 1988; Günther, 2010; Hofmann and Baumann, 1984; Staude et al., 2009; Staude et al., 2011), or the Tertiary due to the Upper Rhine Graben formation (Günther, 2010; Hofmann and Baumann, 1984; Staude et al., 2009; Staude et al., 2011; Werner and Dennert, 2004; Werner and Franzke, 2001; Werner et al., 2002). Veins are of hydrothermal origin causing oxidation in a surrounding contact zone (Günther, 2010). They are associated with extensional fractures that underwent multiphase opening (Günther, 2010; Werner and Dennert, 2004). Those fluid pathways accommodate pore fluids originating e.g. from the graben center (Werner et al., 2002), which are enriched in Na, Ca, and Cl (Dill, 1988; Günther, 2010). They descend via fractures into the crystalline basement, where Ba- and metal ions are dissolved, and after subsequent heating rise along tectonic structures into the overlying strata (Günther, 2010; Werner et al., 2002).

4.4 Materials and methods

Overall 42 samples representing two lateral profiles, containing two separate lithofacies types, were taken along two perpendicular quarry walls of the lowermost berm (berm 0 after Werner et al., 2013). The sampled rock walls have orientations of 098/90° and 188/90°.

Of the samples, 23 originate from a cross-bedded facies (15 horizontal, 8 vertical plugs), three of those horizontal plugs were retrieved from overlying cross-bed sets for calibration, while 19 samples were derived from wavy or ripple-laminated bedding (12 horizontal, 7 vertical plugs). For all 42 samples, one-inch plugs, which were cut to lengths between 2.5 to 4 cm and oven-dried at 40°C for 7 days, as well as thin sections were won.

Lithofacies classes have been described following Miall (1977, 1996) (Tab. 4.1).

Table 4.1: Lithofacies types present in the study area after Miall (1996).

Facies code	Facies	Sedimentary structures	Interpretation
St	Sand, fine to very coarse, may be pebbly	Solitary or grouped trough cross-beds	Sinuous-crested and linguoid (3D) dunes
Swr	Sand, very fine to coarse	Wavy lamination with subordinate ripple cross-lamination	E.g. ripples (lower flow regime)

For the wavy- to ripple-bedded strata the ripple index has been calculated as the ratio of wave length or crest spacing and height, which corresponds to the double amplitude (Tanner, 1967). It is used to identify the basic depositional setting in which the sediment could form ripples as wind ripples with an index > 15 and as subaqueous ripples with an index < 15 (Tanner, 1967).

Porosity is measured with a semi-automated micromeritics Accupyc II 1340 helium pycnometer. Permeabilities are measured using single-phase gas flow in a semi-automated Fancher type isostatic flow cell (Model Rieckmann) applying 1.2 MPa confining pressure and using compressed air ($\eta = 1.82*10^{-5} - 1.83*10^{-5}$ Pa*s) as permeant. Permeabilities are derived using Darcy's law and applying the Klinkenberg correction for slip flow (Klinkenberg, 1941; Rieckmann, 1970).

Plug caps are impregnated with blue-dyed epoxy resin to prepare thin sections to a thickness of 30 µm. A combination of Alizarin Red S and potassium ferricyanide in 0.5 % HCl was used to aid identification of carbonate cements in two thin sections.

The mean grain size is calculated from measurements of the long axes of 100 detrital grains per thin section on a grid adjusted to the largest observed grain size to gain area-weighted results for the apparent grain size in a 2D thin section (Becker et al., 2017; Busch et al., 2018; Schmidt et al., 2020a; Walderhaug, 2000; Walderhaug et al., 2012; Wüstefeld et al., 2017). Grain size classes follow the Udden-Wentworth scale (Udden, 1914; Wentworth, 1922). Sorting was calculated from the measured grain sizes with the formula of Folk and Ward (1957). The grain morphology is classified using the comparator after Füchtbauer (1988c).

The rock composition was determined by point counting samples with 300 steps per thin section on a Leitz Aristomet transmitted light microscope equipped with a semi-automatic Pelcon Point counting stage. The step length is adjusted to the maximum grain size. Mineral accumulations of iron oxide (grains, iron-rich pseudomatrix, pigments, pore-lining and filling iron oxide) are considered as detrital constituents, as ductile deformation during compaction renders any differentiation between authigenic and detrital iron oxide impossible. The detrital composition is plotted after Folk (1980). Grain coating coverages of hematite and illite are determined on 50 grains per sample (Busch et al., 2017; Taylor et al., 2004). Only the surface fraction of a grain in contact with the intergranular volume after Paxton et al. (2002) is considered for the analysis. Individual values range from no grain coating (0 %) to complete grain coating (100 %) in steps of 10 %. Minor hematite pigmentation of a grain is registered with a grain coating coverage of 5 %. A sample's grain coating coverage results from the average value of the 50 grains.

Compaction can be assessed by quantifying the observed types of grain-grain contacts as tangential or point (P-), long (L-), concavo-convex (C-) or sutured (S-) contacts following

Taylor (1950), which reflect increasing degrees of compaction. Grains featuring no contacts are classified as floating grains (F-contacts). The degree of compaction is quantified by point counting those five categories among 100 grains in all samples. Only the contact displaying the maximum grade of compaction was counted per grain. The intergranular volume (IGV) as defined by Paxton et al. (2002) is the sum of intergranular pore space, intergranular cement volumes, and depositional matrix indicating compaction. The calculation of the minus-cement porosity (or intergranular volume after Lundegard, 1992) can be used to assess the compactional porosity loss (COPL) and cementational porosity loss (CEPL) (Lundegard, 1992). For those parameters an initial porosity of 45 % for un-consolidated sand in a fluvial depositional setting is assumed after Wilson and McBride (1988).

The spectral gamma ray signal was measured at all sampling spots of the lateral profiles of the two lithofacies types with a hand-held, auto-stabilizing RS-230 BGO Super-SPEC spectrometer. The decays of the elements potassium, uranium, and thorium are detected over a period of 3 minutes and elemental contents are converted into API-values (Rider and Kennedy, 2011) (Eq. 4.1).

$$SGR(API) = 16.32API \times K(\%) + 8.09API \times U(ppm) + 3.93API \times Th(ppm) \tag{4.1}$$

4.5 Results

4.5.1 Outcrop description

The outcrop features approximately 30 m of reddish-brown sandstone beds that are produced on 4 different berms. The assigned lithofacies types are St and Swr based on the classification after Miall (1996). The first, St, is a medium- to very coarse-grained sandstone with optional pebbles deposited as trough cross-beds via dune migration (Miall, 1996). Since ripple laminae in outcrop are on average less than 2.5 cm thick, which is the diameter of plug samples, a new combined class is introduced to appropriately represent wavy and ripple beds co-existing in the plug samples (Swr).

The quarry walls between berms 0 and 1 are approximately 10 m high and feature thick-bedded to massive layers of 0.7 to 2.0 m thickness with primarily subhorizontal erosive contacts (Fig. 4.3). Bedding contacts sometimes coincide with laminated to platy claystone layers that continue for several meters before pinching out.

The basal, on average 1.2 m thick layer constitutes a trough cross-bedded, reddish-brown sandstone coset (lithofacies type St). The troughs are visible along the N-S striking wall and show basal concave-up, erosive surfaces (Fig. 4.3b) with an apparent dip of tangential foresets to the East along the W-E striking wall. The sets have varying thicknesses of 6 – 50 cm.

The bed in the hanging wall is on average 80 cm thick and consists of a sandstone with wavy, horizontal, sometimes parallel bedding that is subordinately alternating with ripple structures (Swr, Fig. 4.3a, c). Where sampled, those waves and ripples are symmetric and in phase with a wave-length of $\lambda = 2 - 5$ cm and an amplitude of ≤ 0.5 cm (Fig. 4.3c). Climbing ripples exhibiting a phase shift also occur throughout the outcrop. Assuming an average wave length of 3.5 cm, a ripple index of 7 can be calculated. Other descriptive features are the interfering of otherwise straight ripple crests visible on mined blocks (Fig. 4.3c). The wavy-bedded unit shows dark reddish-brown laminae with a thickness of 1 mm and a higher silt- and clay content (Fig. 4.3a).

Other features are reddish-yellow alteration spots occurring in both lithofacies types. They coincide with dark brown to black spots of a negative relief within the rock wall. Occasionally bleached horizons or spots are observed on the laminae-scale in both lithofacies types (Fig. 4.3a). Similarly, dark reddish-brown clayey to silty, angular rip-up clasts are present in both units but reach lengths of ≤ 13 cm in the cross-bedded unit compared to few centimeters in the wavy-bedded unit.

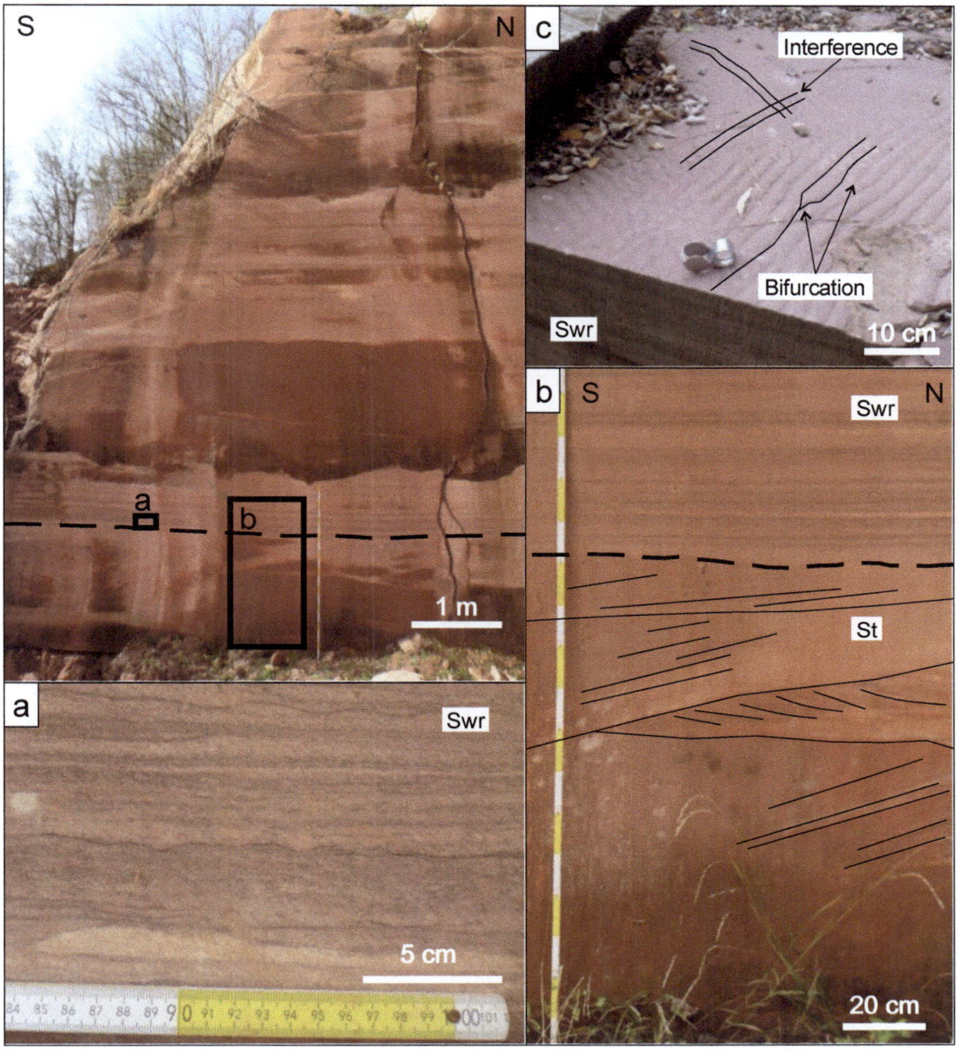

Figure 4.3: Photograph of the N-S striking outcrop wall. The dashed line represents the contact between the cross-bedded foot wall (St) and the wavy-bedded hanging wall layer (Swr). a) Close-up of the Swr lithofacies type displaying dark reddish-brown laminae and bleached yellowish-white zones. The wavy bedding occurs horizontally and (sub)parallel and appears to be often symmetric in cross-section. b) Close-up of the cross-bedded lithofacies type (St) with bimodal apparent dip directions. The hanging wall above the dashed line constitutes wavy beds. c) Mined block in the quarry displaying ripples of the wavy-bedded facies visible in cross-section in the quarry walls (Fig. 4.3a). The ripples show bifurcations of ripple crests and interference of two different wave directions.

4.5.2 Porosity and permeability

Measurements yield porosity values of 1.5 to 21.1 % for the studied samples (Tab. A.2). Permeabilities range from 0.002 to 830 mD (Fig. 4.4, Tab. A.2). While the cross-bedded samples (St) contain on average 17.0 % porosity (max. 21.1 %, min. 12.2 %), the wavy-bedded layers (Swr) have an average porosity of 10.1 % (max. 14.6 %, min. 1.5 %). For the lithofacies type St an average permeability of 236 mD is achieved (max. 830 mD, min. 0.002 mD), for lithofacies type Swr permeabilities are on average 0.09 mD (max. 1 mD, min. 0.002 mD). The outlier of 0.002 mD for the cross-bedded lithofacies is a vertical plug ending in a claystone horizon or large rip-up clast (LK-34). Neglecting that sample, the minimum permeability for the cross-bedded lithofacies type is 0.1 mD and leads to a range of 10^{-1} to 10^{2} mD for sandstone cross-beds, whereas in the wavy strata values range from 10^{-3} to 10^{0} mD.

Vertical plug samples (Δ) were measured in addition to the horizontal plugs of two directions (N-S striking plug × and W-E striking plug +) to get an overview of the 3D permeability anisotropy (Fig. 4.4). In the cross-bedded facies lowest permeabilities of approx. 10^{-1} mD, neglecting the outlier LK-34, are present in vertical plug samples, while highest permeabilities of 10^{2} mD in that lithofacies type may be measured in any of the three sampled directions. For the wavy-bedded lithofacies lowest permeabilities of 10^{-3} mD are registered in any of the three directions, while highest permeabilities of 10^{-1} to 10^{0} mD are only achieved in plugs striking N-S (×).

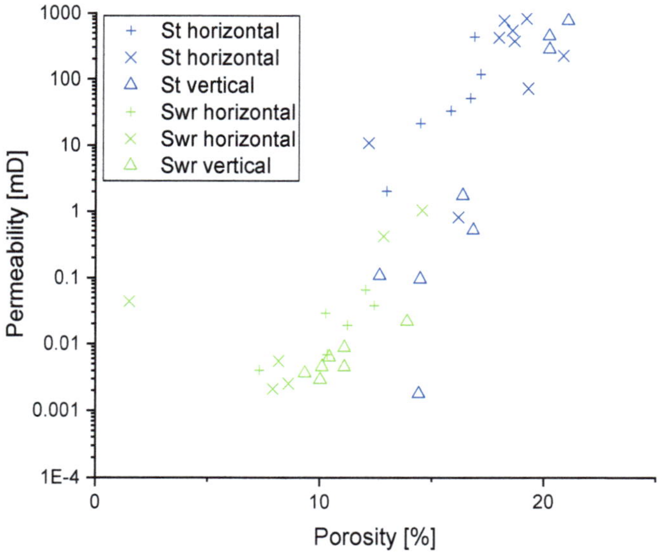

Figure 4.4: Figure caption on the next page.

Figure 4.4: Porosity-permeability-plot for all samples of the study area. Generally, the lithofacies type St (blue) displays more favorable reservoir properties than Swr (green). For each lithofacies type, horizontal plugs striking N-S (×) and W-E (+) and vertical plugs (Δ) show mostly similar values. In St samples lowest permeability is measured in vertical plugs, while in Swr samples the highest permeability is registered for two N-S striking samples.

4.5.3 Texture and gamma ray measurements

The studied outcrop features very fine- (N = 1), fine- (N = 25) to medium-grained (N = 16) sandstones. The macroscopic observation that more fine-grained, dark material is incorporated in the lithofacies type Swr is proven by an average grain size of 0.22 mm (max. 0.31 mm, min. 0.13 mm) (Fig. 4.5a). Lithofacies type St samples are medium-grained and show an average grain size of 0.27 mm (max. 0.33 mm, min. 0.18 mm) (Fig. 4.5a). Both lithofacies types show an overall moderate sorting except for cross-bedded samples LK-18 and -39, which are well sorted.

Average spectral gamma ray values of 86.7 API (max. 95.8 API, min. 80.0 API, Fig. 4.5b) for the Swr lithofacies type are higher than for the St lithofacies type at 66.8 API (max. 83.6 API, min. 57.0 API).

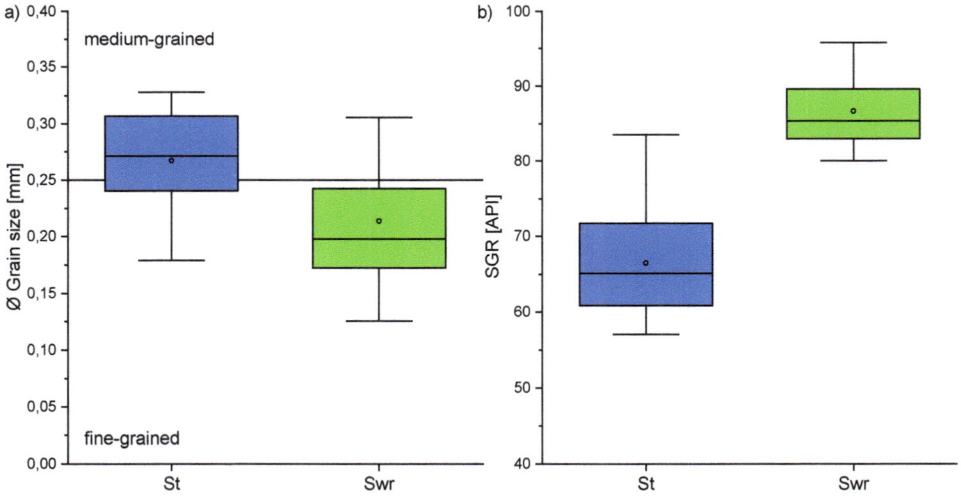

Figure 4.5: a) Box plot of the mean grain size distribution in samples of the St (blue) and Swr (green) lithofacies types. b) Box plot of spectral gamma ray ranges in API of the two lithofacies types with overall higher values in samples from Swr samples.

4.5.4 Petrography

Detrital composition

The samples from the study area display a homogenous mineral composition with the majority of samples being subarkoses of high compositional maturity (Fig. 4.6). This is valid for both, St and Swr lithofacies types. Only three samples are lithic arkoses with a normalized quartz content of less than 75 %. Quantitatively, the overall detrital compositions resemble one another with lithofacies type St consisting of on average 55.8 % quartz grains (max. 65.7 %, min. 43.7 %), 8.0 % feldspars (max. 12.7 %, min. 5.0 %), and 3.7 % rock fragments (max. 7.3 %, min. 1.3 %). Lithofacies type Swr contains on average 57.4 % quartz (max. 64.7 %, min. 44.3 %), 9.2 % feldspar (max. 11.3 %, min. 7.3 %), and 4.2 % rock fragments (max. 7.0 %, min. 2.3 %).

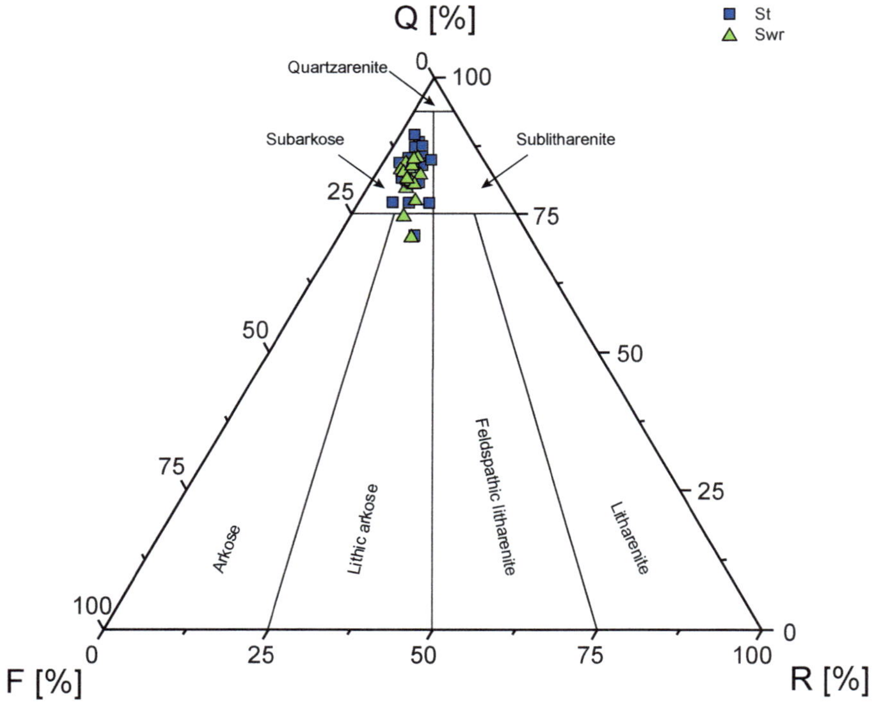

Figure 4.6: Normalized detrital composition of samples from the study area classified after Folk (1980). The majority of samples is classified as subarkoses (N = 39), while three samples are classified as lithic arkoses.

Additionally, the rock samples contain up to 2.7 % of micas (avg. 0.4 %, min. 0 %) and abundant iron oxide-rich phases with on average 11 % (max. 22.7 %, min. 3.7 %). Iron oxide phases are less common in the St lithofacies type with on average 7.9 % (max. 11.7 %, min. 3.7 %), whereas the Swr type contains 14.8 % (max. 22.7 %, min. 6.0 %). Iron oxides appear as pigments or tangential grain coatings (Fig. 4.7a, b, d, 4.8b-d, f, g) as well as a pore-filling phase (Fig. 4.7b). The pore-filling phase may be reddish-brown and homogenously distributed across and within the pore space (Fig. 4.7b) but there are also singular occurrences of dark brown color (Fig. 4.8g, h) that are localized in patches. They coincide with the previously described dark brown to black spots observed macroscopically in samples and the rock wall. Those often cover the grain outline but also reach into the pore space exhibiting a geometric to angular shape (Fig. 4.8h). Additionally, they coincide with abundant intergranular porosity (Fig. 4.8g, h). Accessories are tourmaline and opaque and heavy mineral grains (avg. 0.2 %, max. 1.0 %, min. 0 %). Interspersed, detrital clay is very scarce (avg. 0.06 %, max. 0.7 %, min. 0 %). Rock fragments are sand-, siltstone, and shale rock fragments, undifferentiated metamorphic rock fragments and phyllites, cherts, plutonic, and silicic or feld-spathic volcanic rock fragments. Shale rock fragments were only encountered in samples LK-20 and -23 from the Swr lithofacies type.

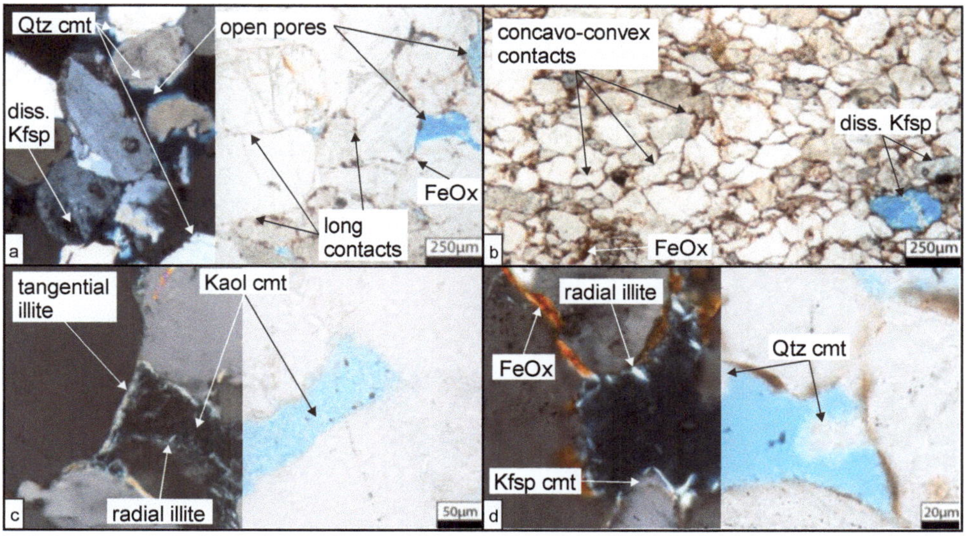

Figure 4.7: Figure caption on the next page.

Figure 4.7: a) Overview of the St lithofacies type displaying open porosity in blue in plain-polarized light (right), primarily long grain contacts, abundant quartz cementation (Qtz cmt) as well as iron oxide (FeOx) as grain-rimming pigments or grain coatings, and a partially dissolved K-feldspar clast (diss. Kfsp, LK-05, left = cross-polarized light/xpl, right = plain-polarized light/ppl). b) Overview of the Swr lithofacies type in ppl with the only visible pore space resulting from the dissolution of K-feldspar grains. Many concavo-convex grain contacts and abundant iron oxide (pore-lining and filling) are observed but rarely quartz cements (LK-21). c) Pore-filling kaolinite booklets (Kaol cmt) with pore-lining tangential and radial illite, the latter partially growing on tangential illite (LK-20, Swr, left = xpl, right = ppl). d) Feldspar cement (Kfsp cmt) in contact with illite. The open pore is lined by iron oxide and radial illite grain coatings and is partially filled by quartz outgrowths (LK-20, Swr, left = xpl, right = ppl).

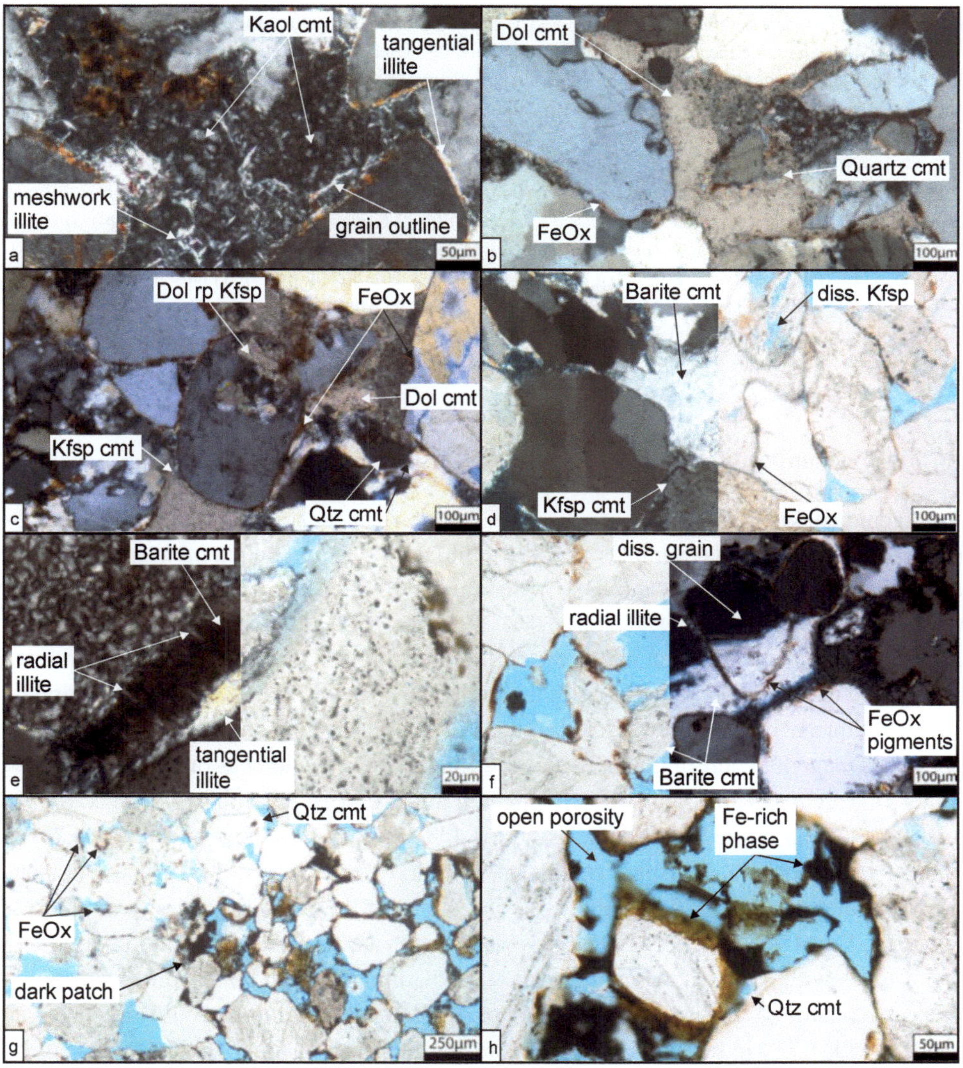

Figure 4.8: Figure caption on the next page.

Figure 4.8: a) Pore space filled by kaolinite booklets including meshwork illite. Tangential illite is present and occasionally appears to delineate the outline of former, now dissolved detrital grains (LK-43, Swr, xpl). b) Pore-filling dolomite (Dol cmt) in the only sample with quantified amounts of dolomite encasing pore-lining iron oxide and quartz overgrowths (LK-46, Swr, xpl). c) Pore-filling dolomite cement surrounding iron oxide grain coatings, quartz cement, and euhedral K-feldspar cement. Dolomite also replaces K-feldspars (Dol rp Kfsp, LK-46, Swr, xpl). d) Pore-filling barite cement (Barite cmt) encasing FeOx-pigmented grains. The sample exhibits minor amounts of syntaxial K-feldspar cement and K-feldspar dissolution (LK-40, Swr, left = xpl, right = ppl). e) Barite cement filling a pore that is lined by radial and tangential illite (LK-40, Swr, left = xpl, right = ppl). f) Barite cement filling an intragranular pore (diss. grain) that still exhibits pore-lining iron oxide (FeOx pigments, LK-40, Swr, left = ppl, right = xpl). g) Overview of an accumulation of a dark brown, iron-rich phase in ppl. The sample yields FeOx-pigmented dust rims and abundant quartz overgrowths whereas underneath the dark brown residue barely any quartz cement formed (LK-19, St). h) Close-up of the iron-rich residue on the detrital grain with minor quartz cement where no residue covers the detrital grain (LK-19, St, ppl).

Authigenic composition

Authigenic phases in the study area are illite, quartz, K-feldspar (Fig. 4.9), kaolinite, dolomite, and barite cements. Samples from the lithofacies type St contain on average 10.4 % quartz cement (max. 14.3 %, min. 3.7 %), compared to substantially less quartz cement in type Swr (avg. 2.0 %, max. 4.7 %, min. 0 %, Fig. 4.7a, b, 4.9). Additionally, a distinct lack of quartz cements is observed where the dark brown to black, iron-rich residue is present, as opposed to the surrounding grains in the sample (Fig. 4.8g, h). In such areas anhedral to subhedral quartz cement only formed in gaps in between the residue (Fig. 4.8h). Illite is present as tangential pore-lining or as a pore-filling phase, e.g. either as radial or meshwork illites. Illite cements are slightly less frequent in the St lithofacies with 4.3 % (max. 8.7 %, min. 0.3 %) than in the Swr lithofacies with 5.4 % (max. 9.3 %, min. 3.0 %, Fig. 4.7c, d, 4.8a, e, 4.9). Blocky K-feldspar cements are similarly scarce in both facies with 0.5 % in the St lithofacies (max. 2.0 %, min. 0 %) and 0.2 % in the Swr lithofacies (max. 1.0 %, min. 0 %, Fig. 4.7d, 4.8c, d, 4.9). Kaolinite aggregates are similarly frequent in both lithofacies types with 0.2 % in St samples (max. 0.7 %, min. 0 %) and 0.3 % in Swr samples (max. 1.0 %, min. 0 %, Fig. 4.7c, 4.8a). Minor occurrences of dolomite could only be point counted with 1.7 % in sample LK-46 from the Swr lithofacies (Fig. 4.8b, c). Barite cement was registered in samples LK-11 and LK-40 with 0.3 % and 0.7 %, respectively, and thus in both lithofacies types (Fig. 4.8d-f). Barite and dolomite were observed more frequently but not encountered during point counting.

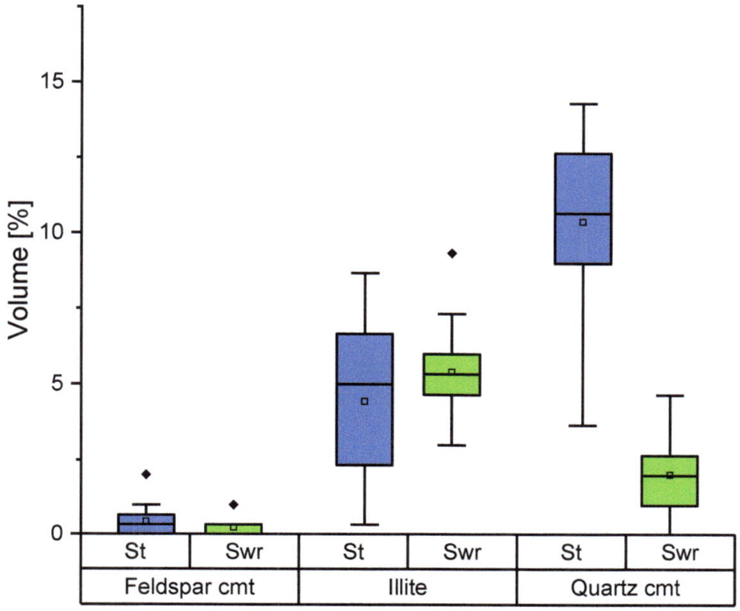

Figure 4.9: Box plot of prominent authigenic mineral phases quartz, illite, and K-feldspar in the cross-bedded (blue) and wavy-bedded (green) lithofacies types.

The different illite morphologies (tangential, radial, and meshwork) appear to exhibit similar textures in both lithofacies. Pore-lining or tangential illites constitute the exception and are more often observed in the Swr lithofacies (Fig. 4.10). It accounts for on average 1.4 % in the St lithofacies (max. 4.3 %, min. 0 %) and 2.5 % in the Swr lithofacies (max. 3.7 %, min. 1.0 %).

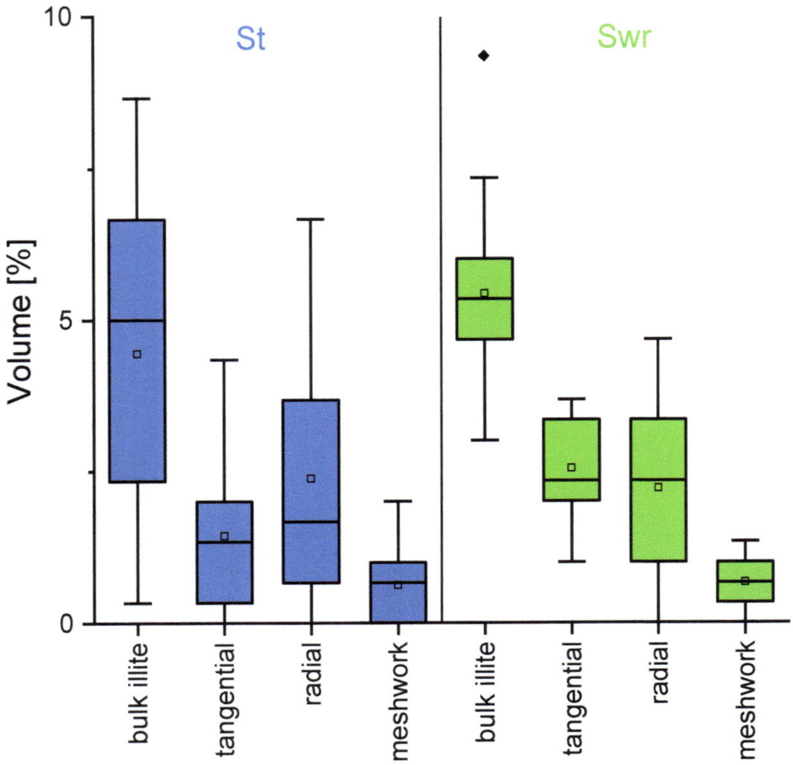

Figure 4.10: Box plot of bulk illite content and all illite cement morphologies present in the St (blue) and Swr (green) lithofacies types.

Mineral replacements by illite and dolomite mainly of detrital feldspars can be observed, with replacive dolomite being present with 0.7 % in sample LK-46 (Fig. 4.8c). Illite replacements are slightly less common in samples from the St lithofacies with on average 1.4 % (max. 3.3 %, min. 0 %). In Swr samples an average of 2.0 % of illite replacements can be point counted (max. 5.3 %, min. 0.3 %). Illite replacements are found along the cleavage of feldspars and within kaolinite booklets and form grids of meshwork illite within the intra- and intergranular pore space, respectively (Fig. 4.8a).

Optical porosity

The St lithofacies contains 11.3 % point counted porosity (max. 15.5 %, min. 6.4 %), while the Swr lithofacies exhibits a point count porosity of on average 8.0 % (max. 13.0 %, min. 2.9 %) (Fig. 4.11, Tab. A.2). The rounded voids containing a dark brown to black coating constitute 0.6 % of the cross-bedded samples (max. 8.0 %, min. 0 %) and 0.7 % of the Swr samples (max. 4.7 %, min. 0 %). A macroscopic assessment of the sampled plugs yields voids in four out of 23 St samples and in three out of 19 Swr samples. Overall intragranular porosity, e.g. in K-feldspars (Fig. 4.7a, b, 4.8d) and rock fragments, from dissolution processes is similar in both lithofacies types (St: avg. 1.3 %, max. 2.7 %, min. 0 %; Swr: avg. 1.2 %, max. 2.3 %, min. 0 %) (Tab. A.2).

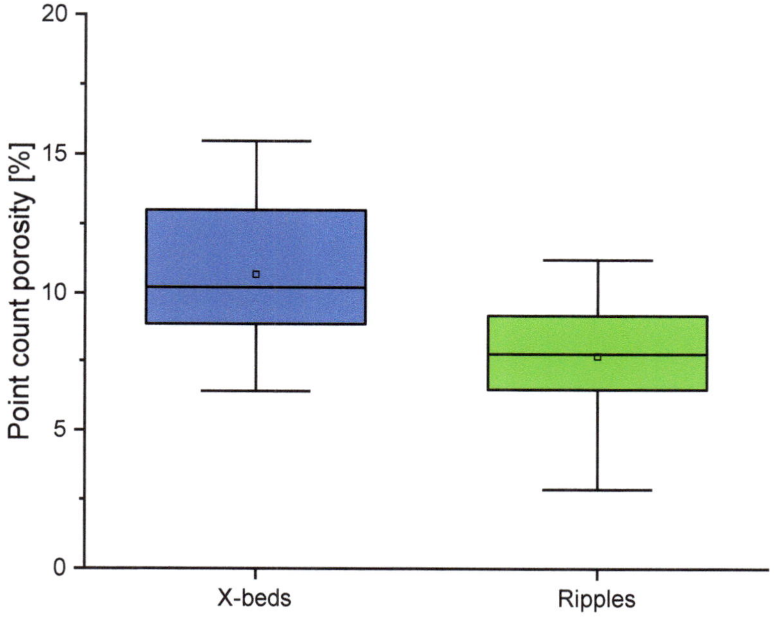

Figure 4.11: Box plot of point counted porosities in samples of the cross- (blue) and wavy- to ripple-bedded (green) lithofacies types.

Grain coating coverages

Grain coating coverages span a range of 25 to 74 % (Fig. 4.12a). St samples exhibit on average 40 % of grain coating coverage (max. 61 %, min. 25 %), while Swr samples have an average grain coating coverage of 67 % (max. 74 %, min. 52 %). Plotting it against the average grain size of a sample indicates that there is no correlation between those two properties (Fig. 4.12b). A cross-plot of quartz cement versus grain coating coverage indicates that samples with a continuous grain coating contain smaller amounts of syntaxial quartz overgrowths (Fig. 4.12c).

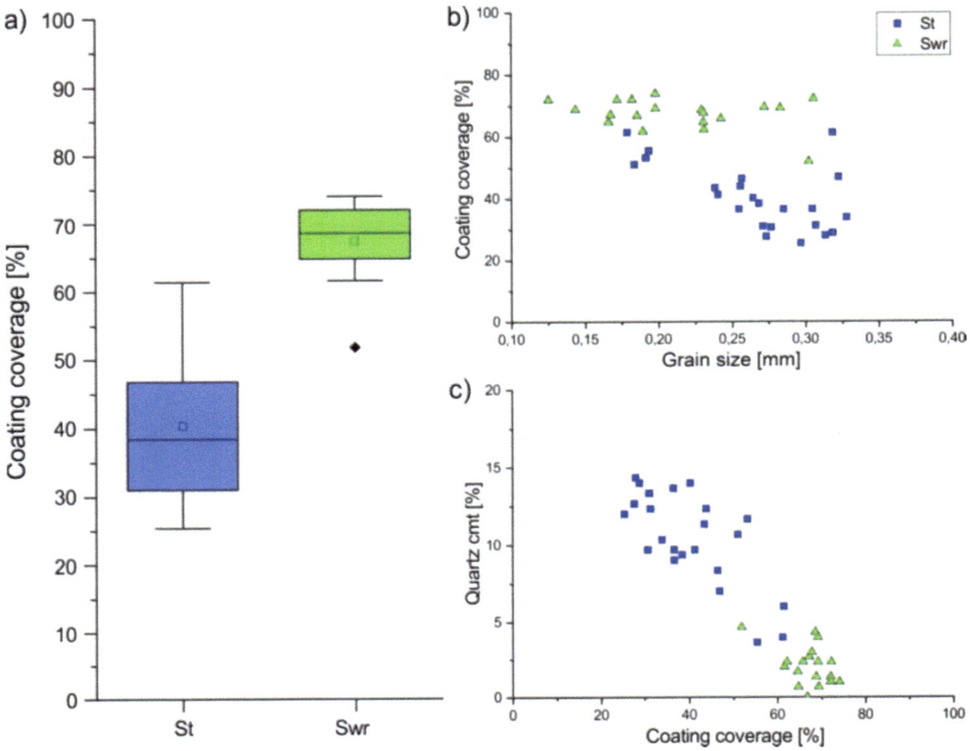

Figure 4.12: a) Box plot of grain coating coverages for St (blue) and Swr (green) lithofacies types displaying a distinct difference in grain coating coverage with St samples showing less well developed clay mineral grain coatings. b) Scatter plot of grain coating coverage vs. grain size displaying no correlation, neither for the Swr nor for the St lithofacies. c) Scatter plot of grain coating coverages correlated with the amount of syntaxial quartz overgrowth cement volumes.

4.5.5 Compaction

Both lithofacies types feature primarily concavo-convex contacts indicating a strong influence of compaction on all sandstone samples. They constitute on average 54 % of grain contacts in the St lithofacies (max. 79 %, min. 34 %) and 88 % in the Swr lithofacies (max. 95 %, min. 78%) (Fig. 4.13). Second most common are long contacts (St: avg. 32 %, max. 43 %, min. 13 %; Swr: avg. 7 %, max. 20 %, min. 1 %). Differences in the frequency of a specific grain contact are not paramount but visible with on average almost half of the grain contacts in the St lithofacies type being point, long or floating (45 %). In the Swr lithofacies type those contacts account for on average 8 %. By contrast, compaction-induced contacts (concavo-convex and sutured) make up almost all contacts in Swr samples (92 %) and slightly more than half of those in St samples (55 %). Grain contacts point toward compaction impacting both lithofacies types but St samples to a lesser extent (Fig. 4.13, Tab. A.2).

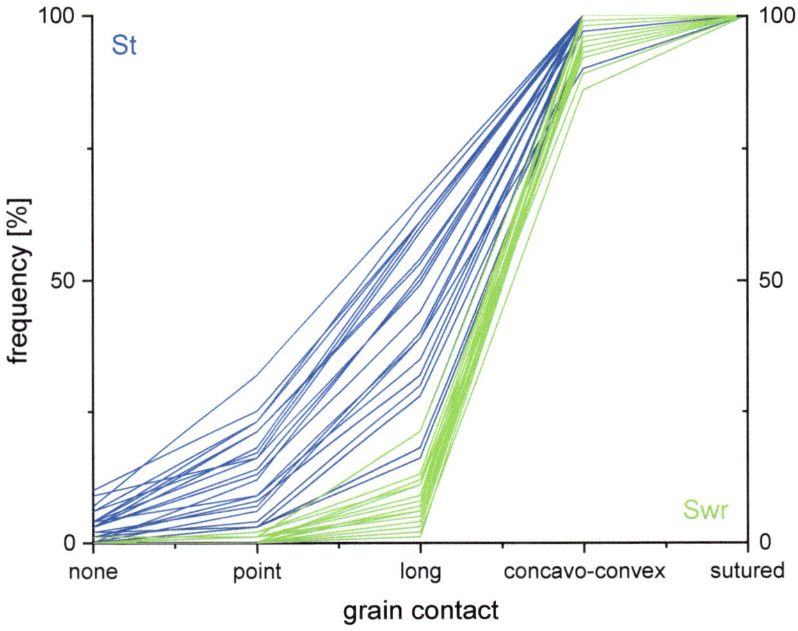

Figure 4.13: Cumulative plot of the distribution of grain contact types in all samples. Both lithofacies types are characterized by primarily concavo-convex grain contacts (blue – cross-bedded, green – wavy- to ripple-bedded).

The IGV evidences differing compaction behavior for the St and Swr lithofacies types (Fig. 4.14). Samples from the St lithofacies have a much higher IGV of 22 % (max. 27 %, min. 13 %) than Swr samples that display an average IGV of 10 % (max. 19 %, min. 5 %). The cause for porosity loss is primarily compaction for both lithofacies types (Fig. 4.14). The average COPL-value of 30 % for the St lithofacies type (max. 37 %, min. 24 %) is substantially lower than the average COPL-value of 39 % for the Swr lithofacies (max. 42 %, min. 32 %). Cementation has a minor influence on porosity losses in all samples (St: avg. 11 %, max. 16 %, min. 7 %, Swr: avg. 5 %, max. 9 %, min. 3 %, Fig. 4.14).

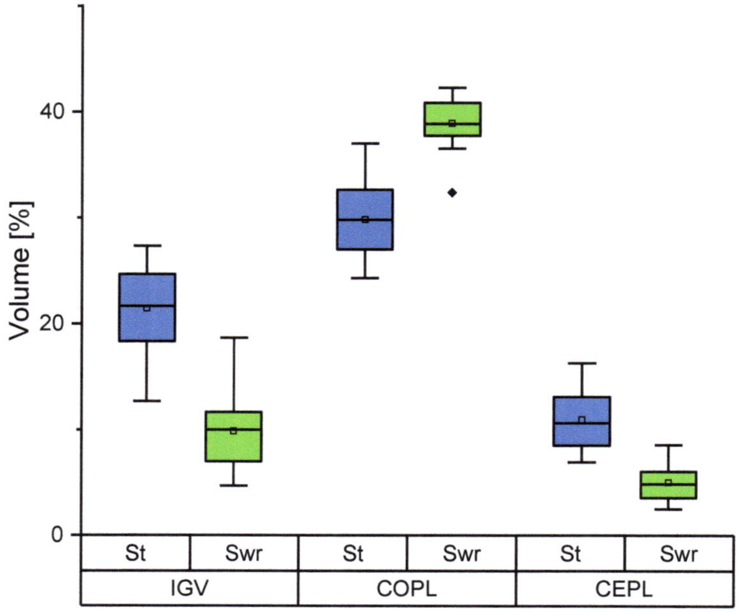

Figure 4.14: Box plot of the ranges of intergranular volumes (IGV) after Paxton et al. (2002) as well as compactional porosity loss (COPL) and cementational porosity loss (CEPL) after Lundegard (1992) for the St and Swr lithofacies types.

Textural properties like grain size and grain coating coverage that are not associated with one another (Fig. 4.12b) are both affecting compaction behavior (Fig. 4.15a, b). Color-coding for grain size illustrates a negative correlation implying that more coarse-grained material is often less compacted but more strongly cemented (Fig. 4.15a). Overall grain coating coverages are positively correlated with compaction and negatively with cementation (Fig. 4.15b). Samples containing larger volumes of quartz cement show smaller COPL-values in both lithofacies types (Fig. 4.15c) while coinciding with higher CEPL-values in those samples. Tangential illite, which is the illite morphology exhibiting the most distinct association with the lithofacies type (Fig. 4.10), is neither controlling the compaction behavior nor the cementation of the two sample types as similar degrees of compaction and cementation may be achieved regardless of high or low tangential illite content (Fig. 4.15d).

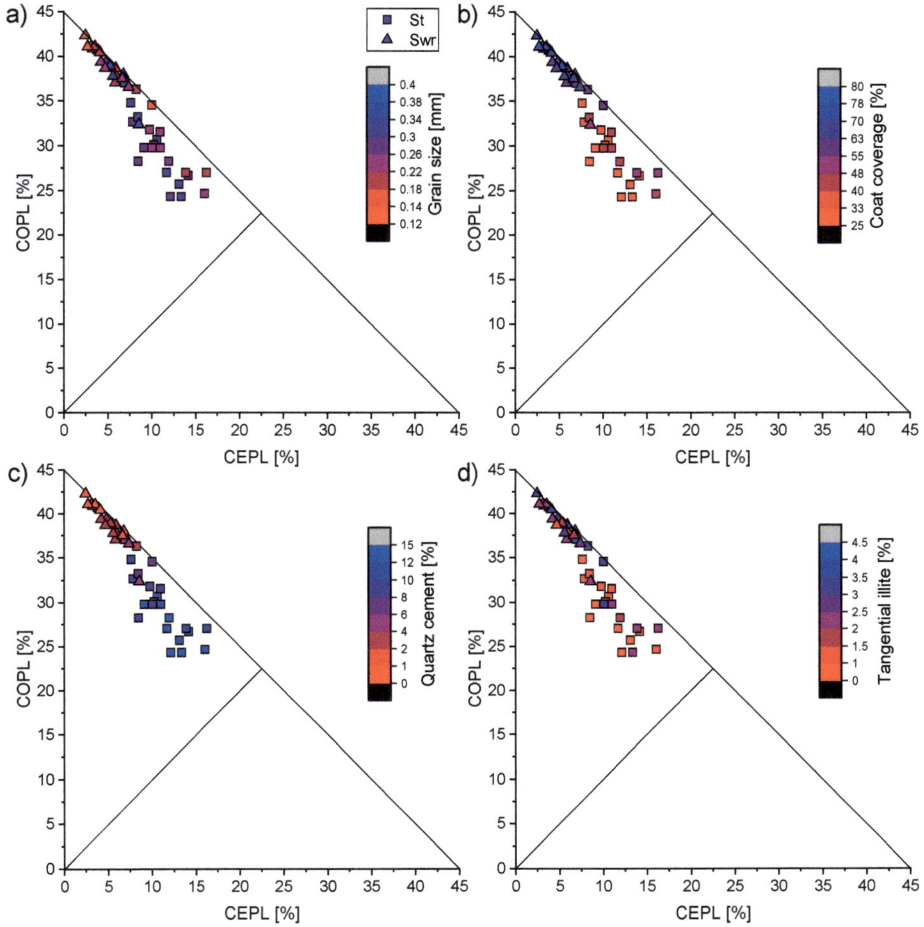

Figure 4.15: Cross-plots after Lundegard (1992) displaying the effect of rock properties on the compaction behavior of individual samples of both lithofacies types. a) Grain sizes appear to control compaction behavior in Swr samples. For the St lithofacies type no clear correlation can be observed. b) Grain coating coverages display a correlation with compaction behavior in the Swr lithofacies. Its effect is less pronounced in the St lithofacies. c) Quartz cement volumes correlate with compaction in both lithofacies types and increase porosity loss by cementation. d) For tangential illite no clear correlation is outlined apart from Swr samples being characterized by generally higher tangential illite contents and COPL-values.

4.5.6 Reservoir quality correlations

Intergranular volumes (IGV) correlate well with porosity and permeability (Fig. 4.16a), with high IGV samples having better reservoir quality. Conversely, high COPL-values

(Fig. 4.16b) correspond to low porosity and permeability, while high CEPL-values correspond to high porosity and permeability (Fig. 4.16c).

Porosity and permeability of the Buntsandstein unit of the study area have been shown to be dependent on the lithofacies type encountered (Fig. 4.4). St samples preserve more porosity and permeability covering a range of 10^{-1} to 10^{2} mD compared to 10^{-3} to 10^{0} mD in the Swr samples, regardless of the sample's orientation (Fig. 4.4). The sample's grain size does not correlate well with reservoir quality in the study area (Fig. 4.16d). However, grain coating coverages, which often correspond to higher COPL values in the study area (Fig. 4.15b), are linked with porosity and permeability. Low porosity and permeability samples have generally higher grain coating coverages (Fig. 4.16e). Grain coating coverages, which differ strongly depending on the lithofacies type (Fig. 4.12a), appear to control reservoir quality in the study area (Fig. 4.16e). Since grain coating coverages also correlate with quartz cement volumes (Fig. 4.12c), both porosity and permeability are generally increased in samples containing larger quartz cement volumes (Fig. 4.16f).

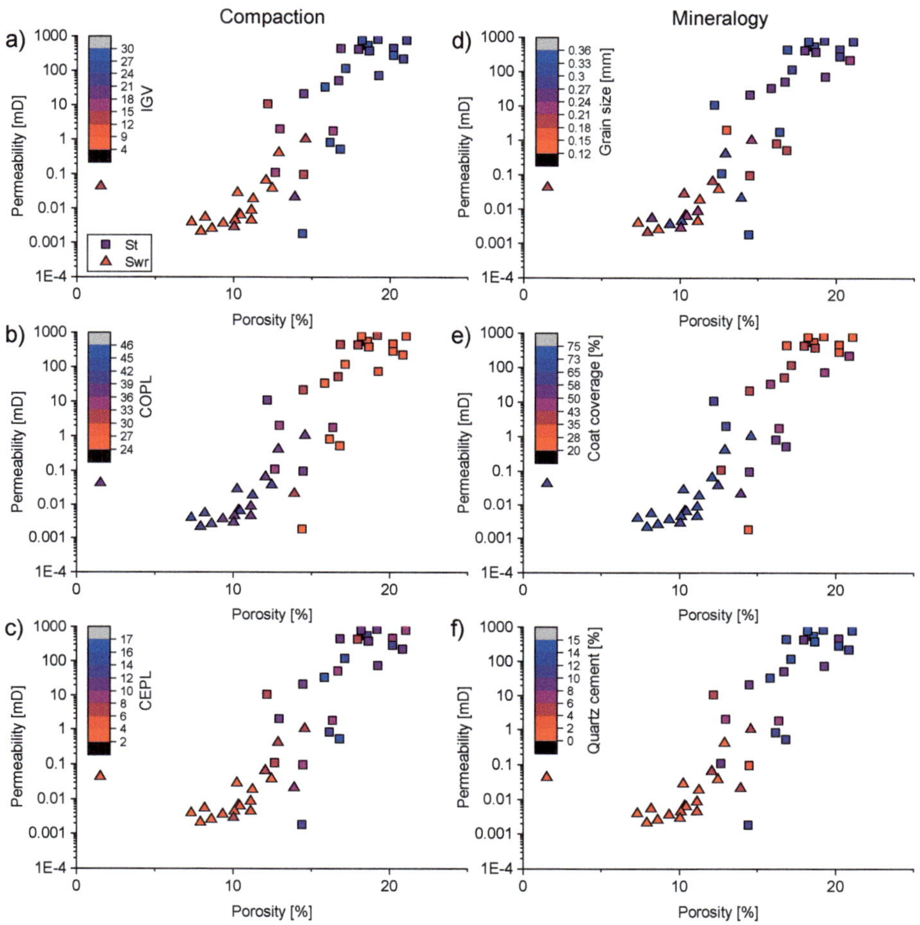

Figure 4.16: Influence of compaction parameters and rock properties on the petrophysical properties of both lithofacies types. a) The intergranular volume is positively linked to petrophysical properties of all samples regardless of lithofacies types except for one outlier. b) COPL correlates negatively with reservoir qualities. c) CEPL exhibits a generally positive correlation with porosity and permeability. d) Grain size does not correlate with reservoir quality. e) Higher grain coating coverages correlate with lower porosities and permeabilities. f) Quartz cements exhibit a positive correlation with porosity and permeability.

4.6 Discussion

4.6.1 Outcrop description and sedimentology

The perpendicular quarry walls allow for the recognition of the cross-beds having been deposited as troughs in the St lithofacies type. The W-E striking wall features a constant apparent dip direction to the East, which is in rough agreement with a documented shedding of the sediment mainly toward the NE during the Buntsandstein (Backhaus, 1974; Günther, 2010; Hagdorn et al., 2009; Röhling and Lepper, 2013; Ziegler, 1990). The N-S striking wall instead shows bimodal apparent dip directions (to the N and S) for the sandstone sets, which have concave-up, erosive bases (Fig. 4.3b). The lithofacies type St as defined by Miall (1996) can thus be clearly identified in the outcrop and is interpreted as sinuous-crested and linguoid dunes linked to the lower flow regime (Miall, 1977a). The second lithofacies type Swr described in this study exhibits mainly wavy bedding in cross-section. A 3D view of a mined block has revealed that the wavy structure often coincides with symmetric, in phase ripple beds (Fig. 4.3c). The lithofacies type Sr proposed by Miall (1996) featuring ripple cross-lamination represents the lower flow regime (Ashley, 1990; Bahlburg and Breitkreuz, 2017; Miall, 1977a, 1996). The derived ripple index of 7 is an indicator for subaqueous ripples. Combining that value with the often high symmetry observed in cross-section (Fig. 4.3a), a generation by wave movement can be interpreted (Tanner, 1967; Tucker, 2003). The interference of some of the straight ripple crests visible on mined blocks (Fig. 4.3c) equally points towards a genesis by wave (or wind) movement, current movement thus being excluded (Tanner, 1967; Tucker, 2003).

Literature proposes braided river systems for SW-Germany during the Lower and Middle Buntsandstein (Backhaus, 1974; Paul, 1982; Werner et al., 2013). The outcrop does not yield any unambiguous evidence for a braided river. Irrespective of this, low stage trough cross-beds in a generally fluvial setting, the relatively low transport energy being indicated also by the comparatively small grain sizes (Fig. 4.5a), are the common sedimentary structure in large scale inclined strata forming fluvial braid and point bars (Bridge, 2006). The erosive horizontal contact can be considered evidence for fluctuations in channel discharge, shifts in channel position or a base-level fall (Bridge, 2006). The latter is ruled out as a low stage ripple-bedded, even more fine-grained lithology follows onto the erosive surface (Fig. 4.3). This implies subsequently lower transport energies and contradicts a base-level fall that would induce higher flow velocities succeeding the initial erosional event (Ashley, 1990). Migration of a braid bar is considered the most probable cause of erosion with the laterally migrating channel center eroding the bar top and ripple beds subsequently being deposited along the outer river bank, where transport energy and water depth are less than in the channel center (Bridge, 2006). By contrast, the succession cross-stratified sandstone – ripple-bedded sandstone as well as the general fining-upward trend from one lithofacies to the other may also represent the typical vertical log through a channel bar deposit (Bridge, 2006). A long-term exposure of the erosive surface is improbable as no vegetation set in, neither did clay from intermittent flooding of an upper-bar deposit accumulate or both were eroded prior to deposition of the ripples. The ubiquitous reddish-brown mud rip-up clasts appear to be intrabasinal, they have been documented to be common, e.g. as channel lag deposits (Henares et al., 2016a). In this study, their frequent occurrence on all stratigraphic levels within a set and in both lithofacies types instead does not yield further information on the depositional setting.

4.6.2 Petrophysical properties and lithofacies type

Porosities for the two lithofacies types show a slight overlap around the overall average porosity of 14 %. Eliminating the outlier, LK-34, the same is valid for permeabilities (Fig. 4.4). Permeability values overlap within a range of 10^{-1} to 10^{0} mD. Otherwise, they occupy distinctly different areas of the porosity-permeability plot (Fig. 4.4). The porosity-permeability plot exhibits one distinct aspect for each of the two lithofacies types: While in the St lithofacies lowest values are present in vertical plug samples, in the Swr lithofacies highest values are measured only in the N-S striking plug samples. Lower permeabilities in the vertical direction of a non-reworked sedimentary rock unit match with the expectation of vertical baffles like normal gradation but potentially also barriers like mud lags or drapes (Allen, 1965; Becker et al., 2019; Becker et al., 2017; Bridge, 2006; Doyle and Sweet, 1995; Ghazi and Mountney, 2009; Hartkamp-Bakker and Donselaar, 2009; Henares et al., 2016a; Hornung and Aigner, 1999; Pranter et al., 2007). The maximum values for Swr samples in N-S direction coincide with along-lamina flow, which is considered to provide favorable flow conditions. Flow across laminae with differing grain sizes would result in lower permeabilities (Hartkamp-Bakker and Donselaar, 2009). Both apparent tendencies of permeability in the two lithofacies types lose in significance when looking at other extrema, e.g. highest permeabilities in the cross-beds. Those are measured in all flow directions, while lowest permeabilities in the wavy bedding also occur in all flow directions (Fig. 4.4). Thus, the porosity-permeability cross-plot shows that neither different horizontal directions nor a vertical direction unambiguously affects petrophysical properties of the different plugs. Permeability across bedding planes is not more strongly impeded than along bedding planes.

4.6.3 Texture and spectral gamma ray

The strong overlap of grain sizes measured in the two lithofacies types (Fig. 4.5a) can be compared to values derived in an overview study on subaqueous, thus fluvial, intertidal, and marine, bedforms by Ashley (1990). Above a threshold of 0.6 m/s at the given average grain sizes, dunes are deposited whereas at less than 0.6 m/s ripple beds form (Ashley, 1990). As both of the lithofacies types are interpreted to coincide with a lower flow regime (Miall, 1977a, 1996), a relatively narrow range of grain sizes scattering mainly around the threshold between fine-grained and medium-grained sandstones matches quite well (avg. $\Phi = 1.9$; threshold $\Phi = 2.0$, Fig. 4.5a, Tab. A.2) (Udden, 1914; Wentworth, 1922). The spectral gamma ray values point toward the general hypothesis of more detrital clay deposition in the Swr lithofacies (Fig. 4.5b), which may be associated with higher point counted tangential illite volumes (Fig. 4.10). Additional textural parameters like sorting and roundness show no remarkable difference between the two sample types.

4.6.4 Paragenetic sequence

In the studied rocks of uniform detrital composition, infiltrated grain coatings composed of clay minerals are considered to be the earliest authigenic phase together with red hematite pigments or continuous hematite grain coatings (Fig. 4.17) (Beyer et al., 2014; Busch et al., 2017; Morad et al., 2010; Walker, 1967a, b, 1975; Walker et al., 1978). Clay mineral grain coatings and hematite pigments directly cover the detrital grains' surfaces and are not preceded by any other diagenetic process or mineral but rather overgrown by all other cement phases (Fig. 4.7a, c, d, 4.8a, e, f) and present at grain contacts.

Grain coating coverages show a clear offset between the two lithofacies types (Fig. 4.12a). They cover an overall broad range of values, which is also common in fluvio-eolian Rotliegend deposits in Europe and Quaternary deposits in the USA (Busch, 2020; Busch et al., 2020). Those previous studies have shown that within one depositional environment grain coating coverage may scatter significantly as well as within one outcrop analog (Busch et al., 2020; Busch et al., 2017). A clear correlation of grain coating coverages with depositional subenvironments of fluvio-eolian sediments and sedimentary rocks could not be established (Busch, 2020; Busch et al., 2020) as opposed to estuarine samples from the UK, where a correlation with depositional subenvironments is derivable (Dowey et al., 2017; Wooldridge et al., 2017, 2019). The significantly greater continuity of clay mineral grain coatings in the Swr lithofacies type is considered to be bound to the transport processes in the river and clay mineral infiltration. Slightly lower transport energies in an overall low flow regime may favor the deposition of a larger clay fraction as indicated by the spectral gamma ray results and tangential illite content for the Swr lithofacies type (Fig. 4.5b, 4.10). The scatter plot with average grain sizes contradicts a commonly proposed correlation between smaller grain size and more abundant grain coatings (Fig. 4.12b) (Wooldridge et al., 2017). This is in agreement with other publications also negating a general correlation of grain coating coverage with grain size (Busch et al., 2020; Dowey et al., 2017).

The dark brown to black pore-filling material occurring in patches probably is the residue of an early pore-filling, nodular cement phase now dissolved (Fig. 4.8g, h, 4.17). Those patches are well known from the Buntsandstein of the Black Forest, especially from the Bausandstein Member ("Wadflecken", Keßler and Leiber, 1994). Nodular carbonate concretions (calcite to dolomite, possibly also siderite or rhodochrosite) present in the subsurface weathered in outcrops leaving behind a brownish to black brown residue of iron- and manganese oxides (Glaser, 1912; Keßler and Leiber, 1994). Consequentially, strong local increases in porosity were observed, which correspond to the results of this study (Fig. 4.8g, h). Angular outlines of the dark brown material may mirror the habit of the pre-existing cement (Fig. 4.8h). The unusual abundance of intergranular pore space and mostly absent concavo-convex grain contacts or syntaxial cements within the patches compared to the surrounding lithology strongly point towards a nodule formation before major compaction, pervasive quartz and illite cementation (Fig. 4.8g, h). An early diagenetic origin of carbonate nodules associated with pedogenesis has also been documented for Buntsandstein sandstones of Central Germany (Beyer et al., 2014) and for Triassic sandstones of the Skagerrak Formation in the British North Sea (Lippmann, 2012).

K-feldspar (and rock fragment) dissolution is interpreted to be an early as well as uplift diagenetic process (Fig. 4.17) prominently reported from sedimentary rocks deposited in semi-arid environments (Filomena and Stollhofen, 2011; Gaupp et al., 1993; Gaupp and Okkerman, 2011; Morad et al., 1989). Where only the outline of a fully dissolved grain is preserved (Fig. 4.7b), dissolution must postdate compaction otherwise the resulting intragranular pore space would most likely undergo deformation (Fig. 4.17). Some intragranular pores are partially filled by other phases along dissolved cleavage planes (Fig. 4.8c). Such early diagenetic dissolution preceding e.g. illite and dolomite provides K^+ in solution for syntaxial K-feldspar cement and illite precipitation (Bjørlykke, 1998; Bjørlykke, 2014; Ehrenberg and Nadeau, 1989; Huang et al., 1986; Lander and Bonnell, 2010; Morad et al., 2010; Thyne et al., 2001). Syntaxial K-feldspar cements occur subhe-dral to euhedral on top of detrital substrate grains (Fig. 4.7d, 4.8c, d, 4.17) overgrowing hematite pigments but not illitized grain coatings additionally indicating an early diagenetic origin as proposed by other studies (Beyer, 2015; Beyer et al., 2014; Füchtbauer, 1967, 1974; Gaupp and Okkerman, 2011; Morad et al., 2010). Due to the small cement volumes, additional relative age relations to other authigenic phases cannot be established for K-feldspar. In addition to dissolution, K-feldspar grains were altered when the sediment was flushed by undersaturated, meteoric waters, which led to the formation of kaolinite booklets during early diagenesis (Fig. 4.7c, 4.8a, 4.17) (Becker et al., 2017; Bjørlykke, 2014; Füchtbauer, 1974; Lanson et al., 2002; Morad et al., 2010; Weber and Lepper, 2002; Worden and Morad, 2003). Their minor volume in the bulk rock of the study area could be due to significant illitization utilizing the available K^+ from feldspar dissolution or from an external source (Busch et al., 2020; Ehrenberg and Nadeau, 1989; Lander and Bonnell, 2010; Lanson et al., 2002). A relatively dry environment with only scarce meteoric waters promoting kaolinitization may have also caused low kaolinite volumes (Becker et al., 2019; Becker et al., 2017).

Considering feldspar dissolution but small cement volumes, it can be assumed that K^+ was available, as were elevated temperatures in order for illitization of kaolinite and of primary clay mineral grain coatings to take place (Bjørkum et al., 1993; Bjørlykke, 1998). Tangential illite covers detrital grains, is present at grain contacts (Fig. 4.7c, 4.8a, e), and is interpreted as a burial diagenetic replacement of early diagenetic infiltrated clay mineral grain coatings, likely of a smectitic composition. This is in agreement with other sandstones deposited under arid conditions (Liewig and Clauer, 2000). The required K^+ ions are likely sourced from K-feldspar and rock fragment dissolution (Fig. 4.17) (Morad et al., 2010). For radial illites, other studies have proposed a need for tangential illite cutans as substrate (Molenaar and Felder, 2018). That observation has not been con-sistently made in this study, where radial illite rims may occur with (Fig. 4.7c) and without any visible tangential substrate (Fig. 4.7d, 4.8e, f). Radial illite may also rim detrital, sometimes hematite pigmented or coated grains that exhibit quartz outgrowths in the direct vicinity where there is no grain coating phase (Fig. 4.7d). A burial diagenetic origin can thus be interpreted for radial illite as proposed in literature (Molenaar and Felder, 2018) predating syntaxial quartz cementation.

A broad scatter of bulk illite volumes as in the St lithofacies (Fig. 4.10) is also known from other Rotliegend eolian red beds as well as Carboniferous fluvial sandstones in N-Germany (Becker et al., 2017; Busch et al., 2018). Tangential illite exhibiting a recognizable discrepancy between the St and Swr lithofacies (Fig. 4.10, Tab. A.2) are considered to result from early clay mineral infiltration (Molenaar and Felder, 2018). A possible explanation for their consistently higher values and thus larger input of detrital clays at deposition of the Swr lithofacies could match the sedimentological interpretation that flow velocities decreased in between the two cosets of different lithofacies. This would allow the deposition of more fine-grained material, i.e. more clay, in the Swr lithofacies.

Silica as a major reactant for quartz cementation typically results from feldspar dissolution, kaolinitization, and illitization, silica dissolution during chemical compaction, or is externally sourced (Beyer et al., 2014; Füchtbauer, 1967; Morad et al., 2010; Walderhaug and Bjorkum, 2003; Weber and Lepper, 2002; Worden and Morad, 2003). Reactive surface area for syntaxial quartz precipitation is only present where the grain surface is not covered by grain coating clay minerals or in contact with other detrital grains (Ajdukiewicz and Lander, 2010; Ajdukiewicz and Larese, 2012; Bloch, 1991; Busch et al., 2020; Busch et al., 2017; Busch et al., 2018; Heald and Larese, 1974; Walderhaug, 1996). Thus the low IGV, high grain coating coverage samples of the Swr lithofacies contain smaller amounts of quartz cement than high IGV, low grain coating coverage samples of the St lithofacies

(Fig. 4.9, 4.12, 4.14). Compaction is thus interpreted to have been initiated prior to quartz cementation (Fig. 4.17). The observation of quartz cements postdating compaction matches findings characterizing it as a burial diagenetic phase necessitating temperatures above 70-75°C (Fig. 4.17) (Lander and Walderhaug, 1999; Morad et al., 2010; Taylor et al., 2010; Walderhaug, 1996, 2000).

Dolomite cements are rarely observed and only once point counted in samples from the study area (LK-46, Fig. 4.8b, c, Tab. A.2). They are one of the possible minerals forming early diagenetic carbonate nodules that appear to have been fully dissolved in the outcrop leaving behind an iron-rich residue (Fig. 4.8g, h, 4.17) (Beyer et al., 2014; Glaser, 1912; Keßler and Leiber, 1994; Lippmann, 2012). The partial replacement of K-feldspars by dolomite is interpreted to occur after early diagenetic feldspar dissolution by meteoric waters (Fig. 4.8c). The presence in pore spaces already hosting quartz cements indicates that dolomite cements syn- to postdate quartz cementation indicating a burial diagenetic origin (Fig. 4.8b). This is in agreement with studies for the German Buntsandstein, especially in Southern Germany (Füchtbauer, 1967, 1974, 1988a).

Barite appearing pore-filling, grain replacive, and overgrowing grains with pigmented hematite rims, K-feldspar and quartz cements, as well as radial and pore-lining illite grain coatings can be interpreted to postdate all of these phases (Fig. 4.8d-f). This indicates a late burial diagenetic or uplift diagenetic origin of the barite cement after dissolution of detrital grains (Fig. 4.17). Barite cement, which has also been observed as mm-thick veins in the outcrop, is assigned a Tertiary age as the closest known barite exposure, the Diersburg-Reichenbach lode, five kilometers East of the outcrop, exhibits a strike matching the opening of the Upper Rhine Graben (Keßler and Leiber, 1994). Thus, it may be considered to be coeval to the Tertiary graben formation.

Due to the strong increase in intergranular pore space and the almost complete absence of other cementing phases around the dark brown to black residue (Fig. 4.8g), dissolution of the nodules is interpreted to be a late, uplift diagenetic process, which set in after major compaction already ceased at near surface conditions (Fig. 4.17). Other authors describe a weathering of patchy carbonate cements leaving behind a brown iron- and manganese-rich residue to be an observation restricted to outcrops and thus uplift diagenesis (Keßler and Leiber, 1994). This is also evidenced by the dark residue now covering the few quartz cements that formed outside of the assumed early diagenetic, pore-filling cement (Fig. 4.8h). Reddish-yellow alteration spots could only be identified in plugs and are considered to be associated with such dissolution of carbonate nodules and the recrystallization of hematite to form iron(hydr)oxides (goethite). Patchy goethite cements are known to result from an exposure to meteoric waters (Eichhubl et al., 2004).

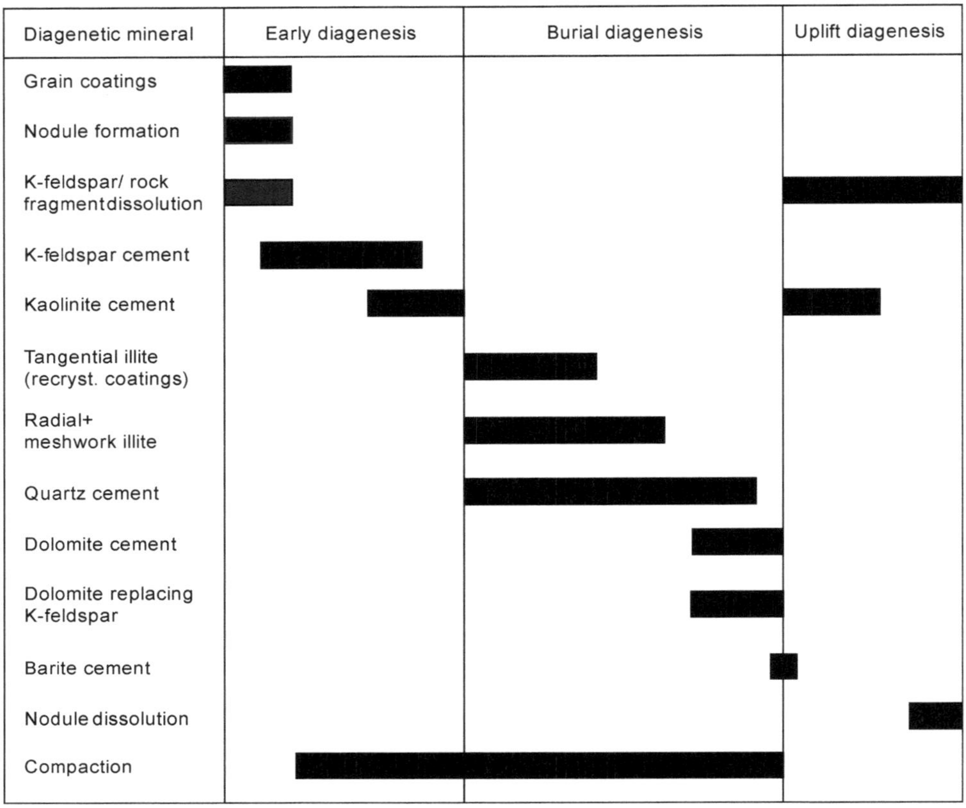

Figure 4.17: Paragenetic sequence for the samples from the study area.

4.6.5 Compaction and reservoir quality

Prevailing concavo-convex grain contacts (Fig. 4.13) and the COPL-values (Fig. 4.14) dominating the samples according to the cross-plot after Lundegard (1992) (Fig. 4.15) underline that the compaction process is of major significance in both lithofacies types of the study area. Still, the very high fraction of compaction-induced grain contacts in the Swr lithofacies (avg. 92 %) compared to the St samples (avg. 55 %, Tab. A.2) implies that chemical compaction of quartz grains in the Swr samples must have been much more pronounced (Bjørkum et al., 1993; Walderhaug, 1996). Much higher remaining IGV- and lower COPL-values as well as higher point count porosity in the St samples support that hypothesis considering that CEPL is not of major importance for porosity loss in neither lithofacies type (St: avg. 11 %, Swr: avg. 5 %, Fig. 4.11, 4.14).

Illitic grain coating coverage, especially at grain contacts, has been linked to a facilitation of chemical compaction (Monsees et al., 2020). While the difference in point counted tangential illite volumes of the two lithofacies types (St: avg. 1.4 %, Swr: avg. 2.5 %, Fig. 4.10) and their impact on compaction and cementation is minor (Fig. 4.15d), SGR-values and grain coating coverages hint at higher clay mineral infiltration in lithofacies type Swr than in St (Fig. 4.5b, Fig. 4.12a). The grain coating coverage's control on compaction behavior in the study area is supported by the COPL-CEPL plot (Fig. 4.15b). Stronger clay infiltration in lithofacies type Swr is in agreement with the sedimentological assumption of lower flow velocities for the Swr lithofacies (Ashley, 1990) coinciding with a larger clay fraction that after illitization could promote chemical compaction at grain contacts. Grain coating coverages have been negatively correlated with quartz cement volumes in this study and by other authors (Fig. 4.12c) (Ajdukiewicz and Lander, 2010; Ajdukiewicz and Larese, 2012; Bloch, 1991; Busch et al., 2020; Busch et al., 2017; Busch et al., 2018; Heald and Larese, 1974; Walderhaug, 1996). As a consequence, quartz cement volumes have been found to exhibit an inverse correlation with compaction (Fig. 4.15c). This means larger quartz overgrowths due to less chemical compaction could have acted as additional framework stabilizing phase.

The relevance of compaction and thus loss of intergranular volume for reservoir quality is visible in the porosity-permeability plots of this study (Fig. 4.16a-c) (Paxton et al., 2002). Grain sizes do not correlate with grain coating coverages (Fig. 4.12b), neither do they directly correlate with reservoir quality (Fig. 4.16d). Instead, increasing average grain coating coverages in lithofacies type Swr (Fig. 4.12) enhance chemical compaction, reduce the intergranular volume (Fig. 4.15b), and diminish reservoir quality (Fig. 4.16e, 4.18). Conversely, where less continuous grain coatings do not promote major chemical compaction in lithofacies type St (Fig. 4.12a, 4.14, 4.15b), substantial quartz overgrowths can fill and preserve the IGV (Fig. 4.15c) and thus porosity and permeability (Fig. 4.16f, 4.18).

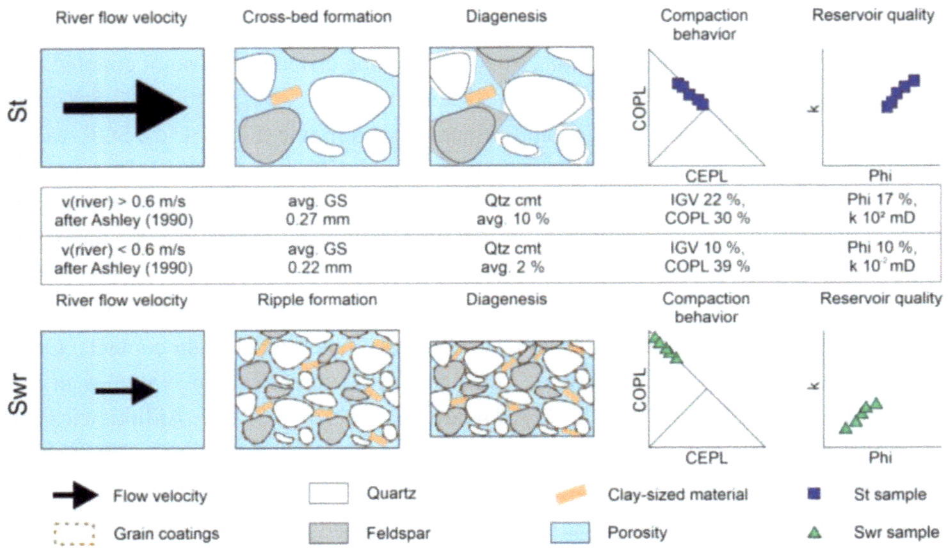

Fig. 4.18: Conceptual model illustrating differing diagenetic evolution paths of the respective lithofacies type as a consequence of minor variations in river flow velocity, thus throughout the depositional process. Grain coating coverage is observed to be proportional to compaction leading to reduced porosity and permeability primarily in lithofacies type Swr.

It has been shown that lithofacies types in the study are linked to reservoir quality (Fig. 4.4) as a result of differing flow velocities during deposition of the respective type (Fig. 4.18). This is in agreement with prior studies that discovered a strong dependency of porosity and permeability on sedimentary processes related to grain size differences and the amount of clay fraction. Those in turn are coupled with the dynamics of the sedimentary processes and transport energy (Hornung and Aigner, 2004). The results from Lahr-Kuhbach do not prove a strong correlation of grain size with lithofacies type nor of grain size and reservoir quality (Fig. 4.5a, 4.16d). Nevertheless, an assumed difference in flow velocity or transport energy derived from the work by Ashley (1990) can be detected, which is responsible for differing clay fractions visible in the SGR readings (Fig. 4.5b) and for the differentiation into two lithotypes.

Similarly, lithofacies associations of braided and meandering river deposits of the Lower Triassic Buntsandstein in SW-Germany show distinctly different permeability values at similar porosities (Wachutka and Aigner, 2001). Braided river deposits with prevailing cross-bedded strata after Miall (1996) are more permeable (St, 10^{-2} to 10^{2} mD) than meandering river deposits of primarily finely laminated or rippled lithofacies types (Sl, Sr, 10^{-3} to 10^{1} mD) (Wachutka and Aigner, 2001). This corresponds to our observations of differing reservoir quality depending on the presence of a cross-bedded or wavy- to ripple-bedded lithofacies (Fig. 4.4). Similar results for lithofacies types after Miall (1996) and reservoir quality originate from Upper Triassic Keuper deposits in Central Germany (Hornung and Aigner, 2004). There, sandy lithofacies types (e.g. Sm, Sh, Srt, St) provide the largest permeabilities of up to 10^3 mD with a peak at 10^2 mD. By contrast, more fine-grained but also sandy lithofacies with ripple cross-bedding (e.g. Fl, Fms, Sr) constitute less permeable samples of a maximum of 10^2 mD and a peak at 10^0 mD (Hornung and Aigner, 2004). This is supported by another study in fluvial Keuper deposits from Germany (Hornung and Aigner, 1999). The sandy lithofacies types St, Sh, Sm, and Sp (parallel lamination) were grouped due to similar permeability distributions exhibiting a peak at 10^3 mD and approx. 23 % porosity. The second subgroup is composed of muddy types Fms, Fl, and Fr (root beds) with a peak at 10^{-1} to 10^0 mD and porosity of approx. 11 % (Hornung and Aigner, 1999). Exceptions were the Sr and Sl lithofacies types, both of which had a broad grain size distribution and porosity and permeability spectra, the latter with poorly defined peaks at $< 10^3$ mD (Hornung and Aigner, 1999). Comparing those two studies (Hornung and Aigner, 1999, 2004) with the results of this study, it can be stated that lithology and grain size do not substantially affect reservoir quality. However, the interplay of lithology and sedimentary structure constituting the lithofacies types (Fig. 4.3) affects reservoir quality (Fig. 4.4, 4.16). The process of grain coating formation and their influence on ensuing diagenetic phases like syntaxial quartz cement (Fig. 4.9, 4.12) and their influence on the compaction behavior (Fig. 4.15) can be interpreted to be the main control on reservoir quality.

4.7 Conclusion

Two lithofacies types of equivalent detrital composition were sampled, one exhibiting trough cross-bedding (St), the other wavy- to ripple-bedded strata (Swr). While the St lithofacies shows a slightly larger average grain size (0.27 mm vs. 0.22 mm), the Swr lithofacies exhibits a stronger spectral gamma ray signal (SGR, 86.7 API vs. 66.8 API). The SGR signal of Swr lithofacies is interpreted to coincide with generally higher clay infiltration at deposition mirrored by a higher grain coating coverage (67 % vs. 40 %) assuming a lower flow velocity for ripple sedimentation compared to cross-beds.

The differences in infiltrated clay minerals also control diagenetic phases resulting in higher illite contents in Swr samples (5.4 % vs. 4.3 %) and major quartz cementation in St samples (10.4 % vs. 2.0 %). Although undergoing same stress conditions, the presence of clay minerals leads to different degrees of compaction in the two lithofacies that govern the precipitation of diagenetic phases. While continuous grain coatings in lithofacies type Swr lead to strong chemical compaction visible in prevailing compaction-induced concavo-convex and sutured grain contacts (avg. 92 %) and reduced IGV (avg. 10 %), IGV is preserved in lithofacies type St (avg. 22 %), where chemical compaction and thus overall compaction is much lower (compaction-induced concavo-convex and sutured grain contacts avg. 55 %). This preserved open porosity available for syntaxial quartz cements and a preservation of optical porosity in St samples (11.3 % vs. 8.0 %). Reservoir quality is controlled by compaction behavior and thus grain coating coverage influencing both chemical compaction and syntaxial cementation.

We conclude that flow conditions at deposition control clay infiltration and allow for a differentiation of lithofacies types. The continuity of the resulting clay grain coatings controls chemical compaction, which in turn steers diagenetic evolution. This way compaction and diagenetic overprint impact reservoir quality leading to better porosity and permeability where less grain coatings/chemical compaction and more quartz cement are encountered. A more profound understanding of the characteristic evolution of intergranular pore space and diagenesis within different lithofacies types could thus lead to a more intuitive mapping of proven, high reservoir quality.

5 Correlation of Buntsandstein reservoir properties

Throughout this thesis two outcrops of Buntsandstein sandstones were investigated: the quarries Röttbach situated North of Wertheim in Lower Franconia, NW-Bavaria (chapters 2 and 3), and Lahr-Kuhbach located North of Freiburg, in the Middle Black Forest (chapter 4). The stratigraphical units encountered in the respective quarries are 1) the Plattensandstein Member of the Röt Formation and 2) the Badische Bausandstein Member of the Vogesensandstein Formation of SW-Germany. The Plattensandstein of the marginal facies of the Germanic Basin (e.g. in the region of outcrop Lahr-Kuhbach) is addressed as a separate formation in SW-Germany, whereas it is classified as a member of the Röt Formation in Lower Franconia (Fig. 5.1).

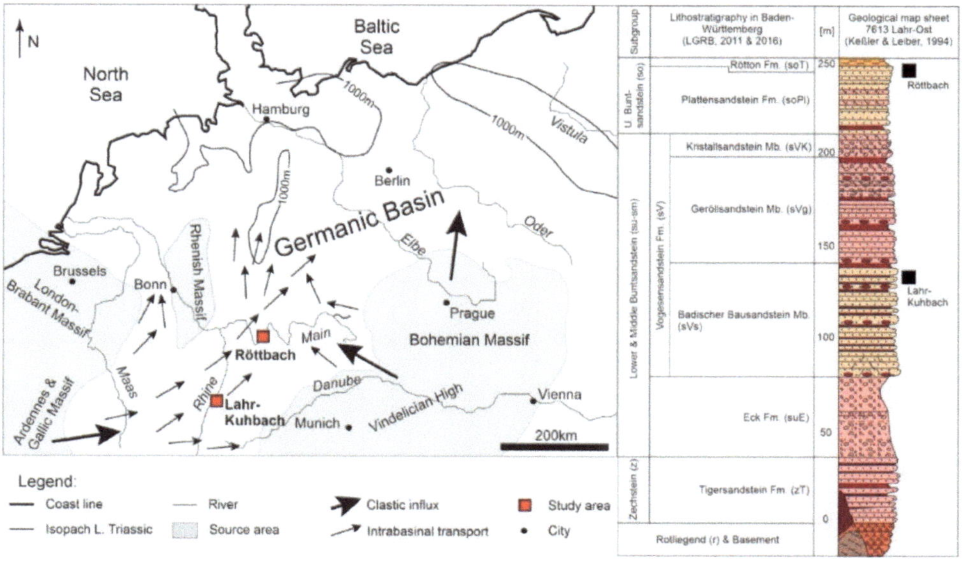

Figure 5.1: Figure caption on the next page.

Figure 5.1: Paleogeographic map of Germany designating the two study areas of this thesis (red squares) within the Germanic Basin, topographic highs (in grey) and shedding directions during the Lower Triassic (arrows, compare Fig. 3.1 after Geyer et al., 2011, Ziegler, 1990). The local stratigraphy for the area of Lahr with both outcrops assigned to their general position within an overall Triassic stratigraphy (black squares). While Lahr-Kuhbach belongs to the Badische Bausandstein Member, Röttbach features the Plattensandstein Member of Lower Franconia.

Both outcrops were deposited during different geological times, with different transport distances and source areas supplying the detritus (Fig. 5.1). The Plattensandstein Member in Röttbach is sourced to a major extent by the Vindelician High approximately 100 km to its SE. The quarry in Lahr-Kuhbach received its detritus mainly from the Ardennes and Gallic Massif approximately 200 km to the WSW (Fig. 5.1). The different source areas and transport distances cause different detrital compositions with Lahr-Kuhbach being of a higher compositional maturity due to prolonged sediment transport (Fig. 5.2). Samples from Röttbach more often plot as e.g. lithic arkoses of lower quartz content and with more feldspars and rock fragments, whose volumetric fractions in Lahr-Kuhbach are diminished due to them decomposing during ongoing transport. Plotting unbleached and bleached Röttbach samples separately in a QFR-plot reveals that bleaching has no decisive impact on detrital composition in the Röttbach quarry and is not associated with depositional processes (Fig. 5.2). The investigated two lithofacies types Swr and St in Lahr-Kuhbach are mostly subarkoses of high compositional maturity. The corresponding QFR-plot does not imply different source areas for the two lithofacies types. The higher maturity of Lahr-Kuhbach samples compared to those from Röttbach is reflected by the QFR-plot, which supports that two source areas or deviating transport distances were relevant for the deposition at the two study sites (Fig. 5.2).

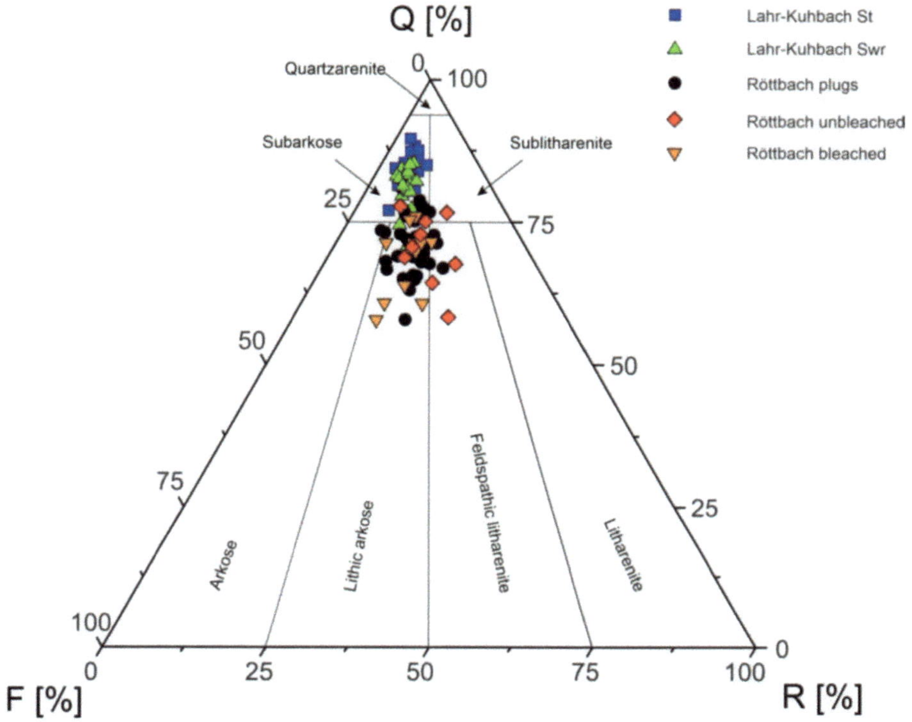

Figure 5.2: Normalized detrital composition of samples from both study areas classified after Folk (1980). The samples are subdivided into the lithofacies types St (blue) and Swr (green) analyzed in Lahr-Kuhbach and the plugs analyzed in Röttbach. The latter are displayed both as bulk analyses (black) and as analyses covering only unbleached (red) or bleached zones (orange) of partially bleached samples.

The analysis of compaction allows a clear differentiation of sample types of one study area. Analyses of the grain contact types revealed differing compaction behavior for 1) unbleached and bleached samples (Fig. 5.3a) and 2) for cross-bedded and ripple-bedded lithofacies types (Fig. 5.3b). The comparison of the two study areas shows a slightly higher degree of compaction for Lahr-Kuhbach samples, which may indicate a deeper burial than experienced by samples in Röttbach. In the cumulative plots the slope of the graph illustrates what contact type is most common. As is shown in chapter 2, the more fine-grained laminae in Röttbach coincide with initially lower porosities, that reduced permeability and thus the effect of bleaching processes. The maintained iron-rich clay mineral grain coatings in those laminae in turn promoted chemical compaction leading to the differing slopes of various grain contacts (Fig. 5.3a) compared to coarse-grained laminae, which after bleaching often developed blocky, diagenetic overgrowths preventing chemical compaction. Regarding lithofacies types in Lahr-Kuhbach, the same correlation can be shown, with higher clay mineral grain coating coverage in the more fine-grained, ripple-bedded layers Swr leading to more pronounced compaction compared to the more coarse-grained, less grain coated, cross-bedded layers St (Fig. 5.3b). In both cases grain contact types as well as the measured IGV mirror this relationship of clay mineral grain coatings being correlated to compaction.

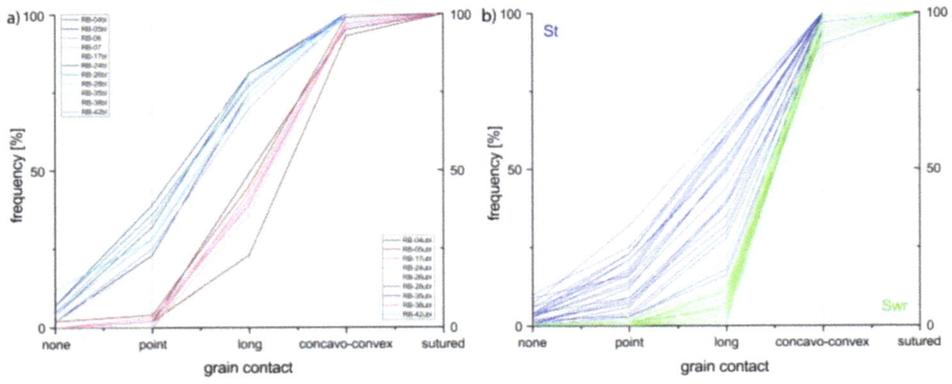

Figure 5.3: Cumulative plots displaying the frequency of the respective grain contact type occurring in a sample. a) Frequency of grain contact types quantified separately for the unbleached (red) and bleached zones (blue) of partially bleached samples of quarry Röttbach (compare Fig. 2.6). b) Frequency of grain contact types as occurring in the lithofacies types Swr (green) and St (blue) of quarry Lahr-Kuhbach (compare Fig. 4.13).

The differing compaction behavior can also be seen in COPL-CEPL plots. They illustrate processes controlling the porosity reduction. Especially the Badische Bausandstein Member in Lahr-Kuhbach lost porosity because of compaction (Fig. 5.4). This effect is more pronounced in the ripple-bedded lithofacies type Swr and linked to its higher clay mineral grain coating coverage. The bulk analysis of Röttbach samples yields a more balanced porosity loss due to combined effects of compaction and cementation (Fig. 5.4). The distinction of unbleached and bleached zones underlines again that compaction is the dominant process in the unbleached zones (Fig. 5.4), as became noticeable when assessing prevailing grain contact types (Fig. 5.3a). Bleached zones instead are characterized by high volumes of blocky cement filling the pore space while stabilizing the grain framework against compaction (Fig. 5.4).

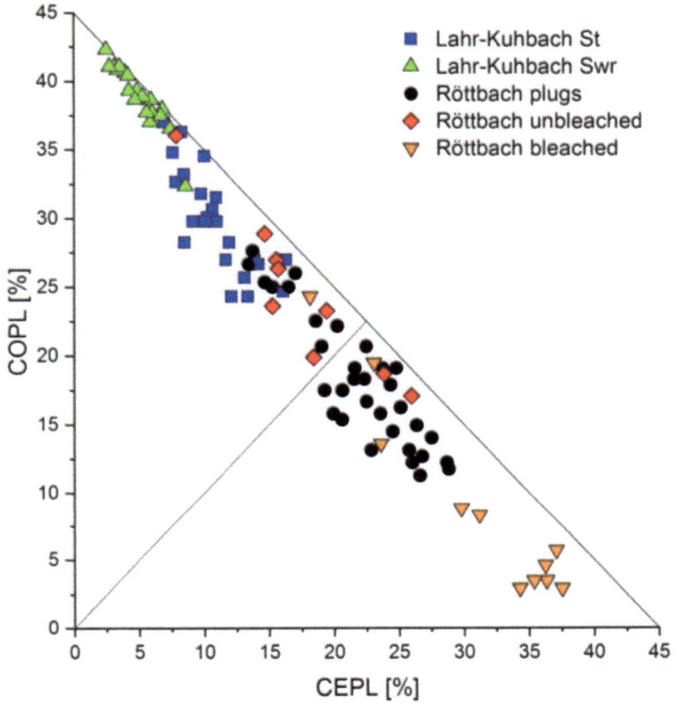

Figure 5.4: COPL-CEPL plot after Lundegard (1992) with all samples from both study areas. Both, the lithofacies type and the presence of bleaching phenomena, are linked to the impact of compaction and cementation. High clay grain coating coverages as observed in lithofacies type Swr and maintained in fine-grained, unbleached laminae facilitate compaction and result in higher COPL-values for the respective study area. More strongly cemented lithofacies type St and bleached laminae could maintain their IGV by stabilizing their grain framework while exhibiting higher CEPL-values.

Plotting permeability against porosity for all plug samples shows a consistent relationship in all Buntsandstein samples of this thesis. The samples of the Badische Bausandstein Member of Lahr-Kuhbach exhibit a broader range of permeability values from 10^{-2} mD to almost 10^{3} mD, while the Plattensandstein Member of Röttbach shows a range from 10^{-1} mD to 10^{1} mD.

Distinguishing the two lithofacies types St and Swr results in more narrow ranges of 10^{-3} mD to 10^{0} mD for the ripple-bedded lithofacies type Swr. The cross-bedded lithofacies type St exhibits a range of 10^{-1} mD to 10^{2} mD. This can be related to minor differences during deposition and the differing clay mineral grain coating coverage, leading to a substantially different evolution during diagenesis. It results in lithofacies type Swr of intermediate reservoir quality and lithofacies type St of good reservoir quality. The latter is owed to grain framework stabilizing quartz cements growing in the absence of continuous clay grain coatings and thus preserving the intergranular volume. Abundant clay mineral grain coatings instead prevented cementation, promoted compaction in lithofacies type Swr and reduced reservoir quality.

The effect of bleaching on permeability could not be measured in this work as samples taken in situ only showed partial bleaching in a single plug. Nevertheless, petrographic analyses revealed that bleaching takes place where initially better reservoir quality / more pathways for fluid flow enhance fluid-rock interactions, which then remove iron-rich grain coatings and enable the formation of e.g. quartz overgrowths. Optical assessment has shown that the grain framework is stabilized and IGV preserved in bleached zones. However, coevally the open pore space is reduced and no longer available for fluid flow. Thus, bleaching appears to be the consequence rather than the cause of higher reservoir quality.

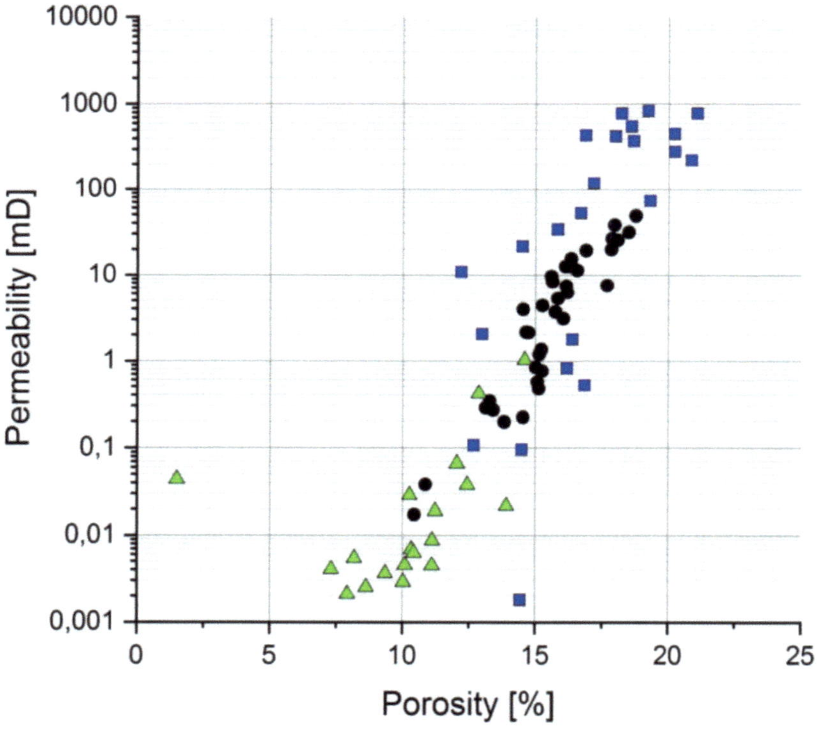

Figure 5.5: Permeability against porosity for all plug samples of Röttbach (black) and Lahr-Kuhbach. In Lahr-Kuhbach plugs were taken from cross-bedded lithofacies St (blue) and ripple-bedded lithofacies Swr (green).

6 Conclusions and outlook

6.1 Conclusions

Bleaching in the studied Lower Triassic Buntsandstein rocks is found along laminae of larger grain size but also known to occur pervasively in the subsurface. Although the bleaching of red sandstones is not necessarily limited to those laminae, it is still controlled by coarse-grained, high porosity laminae and propagates through the surrounding rock from those laminae. It has been shown that a larger average grain diameter of few tens of microns is sufficient to localize the decoloring of the sandy lithology. The larger grain size and, to a lesser extent, the better sorting promote fluid flow along those more permeable horizons causing the decoloring of sandstones, which primarily possess a red stain due to hematite covering the detrital grains.

In the study area in Lower Franconia, Germany, bleached laminae contain almost double the early diagenetic feldspar cement volumes of unbleached laminae. Together with pore-filling dolomite cement, feldspar cements occur euhedral and on top of hematitic grain coatings and thus predate the bleaching process. Quartz cements often overgrow the detrital substrate grains containing fewer hematitic grain coatings. This implies that bleaching fluids reduce hematite causing the discoloration prior to quartz cementation at sufficient burial depths. The overall increased volume of authigenic phases – 11 % more of diagenetic feldspar, dolomite, quartz, and illite – in bleached zones points towards a greater exposure to pore fluids enriched in the ions necessary for mineral precipitation.

Compaction parameters like grain contact types, contact strength, and IGV as well as compactional (COPL) and cementational porosity loss (CEPL) have proven to be powerful tools in assessing porosity reducing processes. Plotting cumulative graphs of the different contact types derived from deposition (floating grains or point contacts) or from compaction during burial (concavo-convex or sutured contacts) represents a hands-on method to derive that the framework stabilizing effect of early to burial diagenetic phases outweighs the extent to which they fill the intergranular pore space. Parameters like COPL and CEPL can be correlated with porosity and permeability to validate if compaction has an impact on reservoir quality.

Framework stabilizing feldspar and dolomite cements, which characterize bleached zones, undergo substantial dissolution during uplift in the studied analog. Point counted porosities in the samples indicate a 3 % higher porosity in the bleached zones. Thus, bleaching has been shown to not necessarily be a process increasing reservoir quality but a viable indicator of primarily good reservoir quality prior to bleaching.

Point bar deposits in a fluvial lithology often represent reservoir units but have been shown to incorporate substantial heterogeneity regarding permeability. In single lateral accretion elements permeabilities vary by three orders of magnitude, exhibiting a coherent trend of deterioration along the channel migration direction. The deterioration of horizontal permeability correlates with increasing amounts of detrital micas and authigenic illite, which are both contributing to mechanical and enhanced chemical compaction. Their increasing presence can be attributed to a decrease in transport energy of the meander, which is the consequence of its increasing curvature.

The presence of compaction enhancing minerals, e.g. micas and clay minerals like illite, is quite well indicated by spectral gamma ray (SGR) readings.

Two sandy lithofacies types in one stratigraphic unit distinguished primarily by their differing sediment structure exhibit marked discrepancies between their respective porosity and permeability values. Those cannot be linked to a major grain size difference, much rather it can be proven that the grain coating coverage by clay minerals is the initial factor causing the ultimately different petrophysical properties.

The higher grain coating coverage in a wavy- to ripple-bedded lithofacies type is responsible for chemical compaction being much more intense destroying large amounts of intergranular volume (IGV) than in the cross-bedded lithofacies type. Those results correspond well with the SGR readings as a proxy for clay mineral content. The grain coating coverage directly controls the amount of syntaxial quartz cement formed during burial, limiting the surface area available for quartz overgrowth precipitation. Those in turn act again as a framework stabilizing phase retaining porosity rather than occluding pores.

The Lundegard plot can reveal, which rock property has a substantial impact on compaction behavior of specific sample sets, e.g. lithofacies types. In the case of lithofacies types St (trough cross-bedded sandstones) and Swr (wavy- to ripple-bedded sandstones), a dependency of the grain coating coverage on compaction can be delineated, as grain coating clays control quartz cementation and promote pressure dissolution along quartz grain boundaries.

Ultimately, the differing sedimentary structures exhibiting very similar grain size ranges, which combined characterize the lithofacies types, result from minute changes in transport energy and thus flow velocity in the river. This reduction in flow velocity causes the transition from cross-bedding to wavy and ripple beds and is also responsible for higher clay contents.

Utilizing compaction parameters and the paragenetic sequences, it can be shown that within a (laminated) lithology of uniform detritus and provenance single rock properties like the average grain size can control a series of rock properties determined after deposition. In a way that processes like compaction and cementation may act differently even on single, mm-sized horizons.

6.2 Outlook

After the bleaching mechanism of fluvial lithologies has been investigated in detail and shown to be reliant on lamination, it is suggested to expand the area under investigation to other sedimentary structures affected by bleaching in other depositional settings like arid or marine deposits. It can be assumed that in overall more coarse-grained rocks of high initial porosity and permeability the localization of bleaching is dependent rather on the density of the reducing pore fluid or on an eventual gradation of the sediment than on grain size-selective lamination. In organic-rich lithologies the reducing agent presumably percolates from the organic particles into the bulk rock regardless of grain size. The controlling parameters therefore would have to be identified for the respective setting.

In Upper Franconia, 150 km East of Röttbach, a second quarry near Trebgast extracts blocks from the Lower Triassic Plattensandstein Member directly overlain by the Röt Claystones. The outcrop contains pervasively bleached sandstones of the same stratigraphy as in Röttbach although the exposed rocks are of a much coarser grain size. They offer a means to study pervasive bleaching and the factors enabling it. As the outcrop is mostly flooded when not mined due to underlying impermeable claystone horizons, the author suggests to research on more abandoned or producing quarries and natural outcrops in the vicinity. The locality in Trebgast is of a second interesting aspect as it is in close proximity (< 5 km) to the Franconian lineament (Fränkische Linie) and surrounded by other minor faults less than a kilometer from the outcrop. If further outcrops with similar bleaching phenomena could be located, it could be investigated if the pervasive bleaching and other rock properties are related to the presence of the faults. If a correlation could be established, it should be attempted to link the degree to which the lithology is bleached and the distance to the next fault. Such an approach could offer new insights on the mobility of (reducing) pore fluids in a faulted sedimentary basin composed of relatively permeable sandstones.

The outcrop analogs studied here are located in Southern Germany. They provide plug samples that were measured in the petrophysical laboratory with respect to porosity and permeability. The conclusion that the effect of bleaching onto reservoir quality is minor should be validated with subsurface samples. In the case of the quarry Lahr-Kuhbach the quality of the analog could be calibrated against samples from the Upper Rhine Graben if taking into account that rock properties may deviate strongly depending on the vicinity to either the western or eastern main graben faults. Among others because of the asymmetric character of the graben and the associated, different maximum burial depths (eastern graben margin more deeply buried) and fluids associated with the specific fault system. At the same time burial depths decrease from NNE to SSW in the graben. Regarding also that sediment transport in the graben area was directed roughly to the North, it becomes necessary to assign an analog from an adequate latitude. Only then it represents conditions at deposition and subsequent burial similar to those of the respective graben samples.

For the Upper Rhine Graben (and southwestern Germany in general) an appropriate burial history would be important because it would enable the construction of kinetic models for quartz and illite cementation. Thereafter, an assessment of authigenic quartz and illite cement volumes in an outcrop analog would allow for an evaluation of its suitability for the respective subsurface target regarding diagenetic phases.

The outcrop in Lahr-Kuhbach features evidence of an early pore-filling carbonate cement, although only one sample contained a quantifiable amount of carbonate cement, which could be identified as dolomite. The dark brown residue interpreted to originate from dissolved iron- and manganese-rich carbonate cement still outlines round patches in the thin sections that point towards formerly cemented nodules. In the samples from Lahr-Kuhbach it should be attempted to make an estimate of the overall nodule volume, their frequency or their volume distribution. By subtracting the grain volume included into a nodule, one could derive the intergranular pore space, which in the subsurface is expected to still be occluded by early diagenetic cement unaffected by uplift diagenesis, i.e. mineral dissolution by meteoric water. Additional subsurface samples from the Upper Rhine Graben could aid in validating those estimates to establish such a calculation for the future use in outcrop analogs containing remnants of nodular cements.

For two lithofacies types a correlation could be established regarding porosity and permeability. A logical follow-up question is whether or not that correlation can be sustained when analysing further outcrops and/or subsurface specimen representing the same lithofacies types. Near Kenzingen in the Black Forest, 15 km South of Lahr-Kuhbach, another quarry produces Middle Buntsandstein rocks, which would offer the opportunity to sample from the same lithology. Trough cross-beds crop out that would be directly comparable with samples in this work. If a stable correlation of single lithofacies types with reservoir quality could be proven to persevere regionally, it might be attractive to compose an atlas of occurring lithofacies types and their petrophysical properties. Thereby reservoir quality prediction for the standard lithofacies associations could be improved significantly if this correlation persists throughout the region.

Regarding grain coating coverage and its influence on permeability, contradictory insights for red beds have been won by authors working in the Lower Permian of the Southern

Permian Basin compared to this work in the Lower Triassic of the Germanic Basin. While a continuous grain coating coverage in the former preserves porosity preventing major quartz cementation, samples in the Buntsandstein appear to undergo a much more pronounced, especially chemical, compaction due to the grain coating clay minerals. Vice versa, quartz cement and additional early pore-filling cements are in the different studies of this thesis the prerequisite for a stabilized grain framework that with-stands compaction in a way that substantial intergranular porosity is maintained. A thorough comparison of paragenetic sequences, burial depths, grain coating coverages, quartz cement volumes, and compaction parameters of the two stratigraphic units, the Permian Rotliegend and the Triassic Buntsandstein, might lead the way toward understanding under which conditions grain coatings act as a risk or as an enabler on reservoir quality.

7 References

Aehnelt, M., U. Hilse, D. Pudlo, K. Heide, and R. Gaupp, 2021, On the origin of bleaching phenomena in red bed sediments of Triassic Buntsandstein deposits in Central Germany: Geochemistry, v. 81, p. 125736.

Aigner, T., and G. H. Bachmann, 1992, Sequence-stratigraphic framework of the German Triassic: Sedimentary Geology, v. 80, p. 115-135.

Ajdukiewicz, J. M., and R. H. Lander, 2010, Sandstone reservoir quality prediction: The state of the art: AAPG Bulletin, v. 94, p. 1083-1091.

Ajdukiewicz, J. M., and R. E. Larese, 2012, How clay grain coats inhibit quartz cement and preserve porosity in deeply buried sandstones: Observations and experiments: AAPG Bulletin, v. 96, p. 2091-2119.

Ajdukiewicz, J. M., P. H. Nicholson, and W. L. Esch, 2010, Prediction of deep reservoir quality using early diagenetic process models in the Jurassic Norphlet Formation, Gulf of Mexico: AAPG Bulletin, v. 94, p. 1189-1227.

Al-Ramadan, K., S. Morad, J. N. Proust, and I. Al-Aasm, 2005, Distribution of diagenetic alterations in siliciclastic shoreface deposits within a sequence stratigraphic framework: Evidence from the Upper Jurassic, Boulonnais, NW France: Journal of Sedimentary Research, v. 75, p. 943-959.

Allen, J. R. L., 1965, Fining-upward cycles in alluvial successions: Geological Journal, v. 4, p. 229-246.

Amelung, W., H.-P. Blume, H. Fleige, R. Horn, E. Kandeler, I. Kögel-Knabner, R. Kretzschmar, K. Stahr, and B.-M. Wilke, 2018, Anorganische Komponenten der Böden - Minerale und Gesteine, Scheffer/Schachtschabel Lehrbuch der Bodenkunde: Berlin, Heidelberg, Springer, p. 11-62.

Anjos, S. M. C., L. F. De Ros, and C. M. A. Silva, 1999, Chlorite authigenesis and porosity preservation in the Upper Cretaceous marine sandstones of the Santos Basin, offshore eastern Brazil, in R. H. Worden, and S. Morad, eds., Clay Mineral Cements in Sandstones: Special publication number 34 of the International Association of Sedimentologists, Blackwell Publishing, p. 289-316.

Ashley, G. M., 1990, Classification of large-scale subaqueous bedforms: A new look at an old problem: Journal of Sedimentary Petrology, v. 60, p. 160-172.

Baaske, U., 1999, Untersuchungen zur Diagenese des Buntsandsteins am Westrand des Rheingrabens (Region Bad Dürkheim/Neustadt a.d.W.): Diploma thesis, Johannes Gutenberg University, Mainz, 142 p.

Backhaus, E., 1974, Limnische und fluviatile Sedimentation im südwestdeutschen Buntsandstein: Geologische Rundschau, v. 63, p. 925-942.

Backhaus, E., 1981, Der marin-brackische Einfluß im Oberen Röt Süddeutschlands: Zeitschrift der Deutschen Geologischen Gesellschaft, v. 132, p. 361-382.

Bahlburg, H., and C. Breitkreuz, 2017, Abtragung, Transport und Ablagerung, Grundlagen der Geologie: Lehrbuch: Berlin, Springer Spektrum, p. 81-93.

Bauer, A., B. Velde, and G. Berger, 1998, Kaolinite transformation in high molar KOH solutions: Applied Geochemistry, v. 13, p. 619-629.

Bauer, A., B. Velde, and R. Gaupp, 2000, Experimental constraints on illite crystal morphology: Clay Minerals, v. 35, p. 587-597.

Beard, D. C., and P. K. Weyl, 1973, Influence of texture on porosity and permeability of unconsolidated sand: AAPG Bulletin, v. 57, p. 349-369.

Becker, I., B. Busch, B. Koehrer, D. Adelmann, and C. Hilgers, 2019, Reservoir quality evolution of Upper Carboniferous (Westphalian) tight gas sandstones, Lower Saxony Basin, NW Germany: Journal of Petroleum Geology, v. 42, p. 371-392.

Becker, I., P. Wüstefeld, B. Koehrer, M. Felder, and C. Hilgers, 2017, Porosity and permeability variations in a tight gas sandstone reservoir analogue, Westphalian D, Lower Saxony Basin, NW Germany: Influence of depositional setting and diagenesis: Journal of Petroleum Geology, v. 40, p. 363-389.

Beitler, B., M. A. Chan, and W. T. Parry, 2003, Bleaching of Jurassic Navajo Sandstone on Colorado Plateau Laramide highs: Evidence of exhumed hydrocarbon supergiants?: Geology, v. 31, p. 1041-1044.

Bertier, P., R. Swennen, D. Lagrou, B. Laenen, and R. Kemps, 2008, Palaeo-climate controlled diagenesis of the Westphalian C & D fluvial sandstones in the Campine Basin (north-east Belgium): Sedimentology, v. 55, p. 1375-1417.

Beyer, D., 2015, Evolution of reservoir properties in the Lower Triassic aquifer sandstones of the Thuringian Syncline in Central Germany: Doctoral thesis, Friedrich-Schiller-Universität, Jena.

Beyer, D., C. Kunkel, M. Aehnelt, D. Pudlo, T. Voigt, G. Nover, and R. Gaupp, 2014, Influence of depositional environment and diagenesis on petrophysical properties of clastic sediments (Buntsandstein of the Thuringian Syncline, Central Germany): Zeitschrift der Deutschen Gesellschaft für Geowissenschaften, v. 165, p. 345-365.

Beyer, D., C. Kunkel, U. Hilse, M. Aehnelt, D. Pudlo, T. Voigt, and R. Gaupp, 2011, Reservoir properties and fluid circulation in Buntsandstein aquifer sandstones of the Thuringian Basin in Central Germany: EGU General Assembly 2011, Vienna.

Bjørkum, P. A., O. Walderhaug, and N. E. Aase, 1993, A model for the effect of illitization on porosity and quartz cementation of sandstones: Journal of Sedimentary Petrology, v. 63, p. 1089-1091.

Bjørlykke, K., 1998, Clay mineral diagenesis in sedimentary basins — a key to the prediction of rock properties. Examples from the North Sea Basin: Clay Minerals, v. 33, p. 15-34.

Bjørlykke, K., 2014, Relationships between depositional environments, burial history and rock properties. Some principal aspects of diagenetic process in sedimentary basins: Sedimentary Geology, v. 301, p. 1-14.

Blanckaert, K., and H. J. de Vriend, 2010, Meander dynamics: A nonlinear model without curvature restrictions for flow in open-channel bends: Journal of Geophysical Research, v. 115.

Bloch, S., 1991, Empirical prediction of porosity and permeability in sandstones: AAPG Bulletin, v. 75, p. 1145-1160.

Bloch, S., R. H. Lander, and L. Bonnell, 2002, Anomalously high porosity and permeability in deeply buried sandstone reservoirs: Origin and predictability: AAPG Bulletin, v. 86.

Bock, H., W. Freudenberger, J. Lepper, P. Schmitt, and J. Weber, 2005, Der Buntsandstein in Main-Tauberfranken (Exkursion B am 31. März 2005): Jahresberichte und Mitteilungen des Oberrheinischen Geologischen Vereins, v. 87, p. 65-96.

Bordy, E. M., and O. Catuneanu, 2002, Sedimentology of the Beaufort-Molteno Karoo fluvial strata in the Tuli Basin, South Africa: South African Journal of Geology, v. 105, p. 51-66.

Böse, M., J. Ehlers, and F. Lehmkuhl, 2018, Deutschlands Norden: Berlin, Heidelberg, Springer, 201 p.

Bridge, J. S., 2006, Fluvial facies models: Recent developments, in H. W. Posamentier, and R. G. Walker, eds., Facies Models Revisited: Special Publication, Society for Sedimentary Geology, p. 85-170.

Brockamp, O., N. Clauer, and M. Zuther, 1994, K-Ar dating of episodic Mesozoic fluid migrations along the fault system of Gernsbach between the Moldanubian and Saxothuringian (Northern Black Forest, Germany): Geologische Rundschau, v. 83, p. 180-185.

Busch, B., 2020, Pilot study on provenance and depositional controls on clay mineral coatings in active fluvio-eolian systems, western USA: Sedimentary Geology, v. 406, p. 105721.

Busch, B., I. Becker, B. Koehrer, D. Adelmann, and C. Hilgers, 2019, Porosity evolution of two Upper Carboniferous tight-gas-fluvial sandstone reservoirs: Impact of fractures and total cement volumes on reservoir quality: Marine and Petroleum Geology, v. 100, p. 376-390.

Busch, B., C. Hilgers, and D. Adelmann, 2020, Reservoir quality controls on Rotliegend fluvio-aeolian wells in Germany and the Netherlands, Southern Permian Basin –

Impact of grain coatings and cements: Marine and Petroleum Geology, v. 112, p. 104075.

Busch, B., C. Hilgers, L. Gronen, and D. Adelmann, 2017, Cementation and structural diagenesis of fluvio-aeolian Rotliegend sandstones, northern England: Journal of the Geological Society, v. 174, p. 855-868.

Busch, B., C. Hilgers, R. H. Lander, L. M. Bonnell, and D. Adelmann, 2018, Reservoir quality and burial model evaluation by kinetic quartz and illite cementation modeling: Case study of Rotliegendes, north Germany: AAPG Bulletin, v. 102, p. 293-307.

Buurman, P., 1980, Palaeosols in the Reading Beds (Paleocene) of Alum Bay, Isle of Wight, U.K: Sedimentology, v. 27, p. 593-606.

Cadle, A. B., and B. Cairncross, 1993, A sandy, bed-load dominated fluvial system deposited by lateral-accretion: Permian Karoo Sequence, South Africa: Sedimentary Geology, v. 85, p. 435-455.

Canham, A. C., M. A. Love, A. Racey, and S. Polachan, 1996, Stratigraphy and reservoir potential of the Mesozoic Khorat Group, NE Thailand: Journal of Petroleum Geology, v. 19, p. 321-338.

Carling, P. A., 1999, Subaqueous gravel dunes: Journal of Sedimentary Research, v. 69, p. 534-545.

Chan, M. A., W. T. Parry, and J. R. Bowman, 2000, Diagenetic hematite and manganese oxides and fault-related fluid flow in Jurassic sandstones, southeastern Utah: AAPG Bulletin, v. 84, p. 1281-1310.

Chang, H. H., 1983, Energy expenditure in curved open channels: Journal of Hydraulic Engineering, v. 109, p. 1012-1022.

Chilingar, G. V., 1964, Relationship between porosity, permeability and grain-size distribution of sands and sandstones: Developments in Sedimentology, v. 1, p. 71-75.

Chu, H., G. Chi, S. Bosman, and C. Card, 2015, Diagenetic and geochemical studies of sandstones from drill core DV10-001 in the Athabasca basin, Canada, and implications for uranium mineralization: Journal of Geochemical Exploration, v. 148, p. 206-230.

Deutsche Stratigraphische Kommission, 2016, Stratigraphische Tabelle von Deutschland 2016: Potsdam, GeoForschungsZentrum.

Dickson, J. A. D., 1966, Carbonate identification and genesis as revealed by staining: Journal of Sedimentary Petrology, v. 36, p. 491-505.

Dill, H., 1988, Geologic setting and age relationship of fluorite-barite mineralization in southern Germany with special reference to the Late Paleozoic unconformity: Mineralium Deposita, v. 23, p. 16-23.

Dong, Y., R. A. Sanford, R. A. Locke, 2nd, I. K. Cann, R. I. Mackie, and B. W. Fouke, 2014, Fe-oxide grain coatings support bacterial Fe-reducing metabolisms in 1.7-2.0 km-deep subsurface quartz arenite sandstone reservoirs of the Illinois Basin (USA): Frontiers in Microbiology, v. 5, p. 511.

Dowey, P. J., R. H. Worden, J. Utley, and D. M. Hodgson, 2017, Sedimentary controls on modern sand grain coat formation: Sedimentary Geology, v. 353, p. 46-63.

Doyle, J. D., and M. L. Sweet, 1995, Three-dimensional distribution of lithofacies, bounding surfaces, porosity, and permeability in a fluvial sandstone - Gypsy Sandstone of northern Oklahoma: AAPG Bulletin, v. 79, p. 70-96.

Eberth, D. A., and A. D. Miall, 1991, Stratigraphy, sedimentology and evolution of a vertebrate-bearing braided to anastomosed fluvial system, Cutler Formation (Permian-Pennsylvanian), north-central New Mexico: Sedimentary Geology, v. 72, p. 225-252.

Ehrenberg, S. N., 1995, Measuring sandstone compaction from modal analyses of thin sections: How to do it and what the results mean: Journal of Sedimentary Research, v. A65, p. 369-379.

Ehrenberg, S. N., and P. H. Nadeau, 1989, Formation of diagenetic illite in sandstones of the Garn Formation, Haltenbanken area, Mid-Norwegian continental shelf: Clay Minerals, v. 24, p. 233-253.

Eichhubl, P., W. L. Taylor, D. D. Pollard, and A. Aydin, 2004, Paleo-fluid flow and deformation in the Aztec Sandstone at the Valley of Fire, Nevada—Evidence for the coupling of hydrogeologic, diagenetic, and tectonic processes: Geological Society of America Bulletin, v. 116, p. 1120-1136.

El-ghali, M. A. K., H. Mansurbeg, S. Morad, I. Al-Aasm, and G. Ajdanlisky, 2006, Distribution of diagenetic alterations in fluvial and paralic deposits within sequence stratigraphic framework: Evidence from the Petrohan Terrigenous Group and the Svidol Formation, Lower Triassic, NW Bulgaria: Sedimentary Geology, v. 190, p. 299-321.

Filomena, C. M., and H. Stollhofen, 2011, Ultrasonic logging across unconformities — outcrop and core logger sonic patterns of the Early Triassic Middle Buntsandstein Hardegsen unconformity, southern Germany: Sedimentary Geology, v. 236, p. 185-196.

Fischer, C., I. Dunkl, H. Von Eynatten, J. R. Wijbrans, and R. Gaupp, 2012, Products and timing of diagenetic processes in Upper Rotliegend sandstones from Bebertal (North German Basin, Parchim Formation, Flechtingen High, Germany): Geological Magazine, v. 149, p. 827-840.

Folk, R. L., 1980, Petrology of sedimentary rocks: Austin, Texas, Hemphill Pub. Co, 182 p.

Folk, R. L., and W. C. Ward, 1957, Brazos river bar: A study in the significance of grain size parameters: Journal of Sedimentary Petrology, v. 27, p. 3-26.

Frank, M., 1937, Ergebnisse neuer Untersuchungen über Fazies und Bildung von Trias und Jura in Südwest-Deutschland: Geologische Rundschau, v. 28, p. 465-498.

Freudenberger, W., 1990, Erläuterungen zu Blatt 6223 Wertheim: Geologische Karte 1 : 25000 von Baden-Württemberg: Stuttgart, Landesvermessungsamt Baden-Württemberg.

Freudenberger, W., G. Geyer, and B. Schröder, 2013, Der Buntsandstein im nördlichen Bayern (nordwestliches Franken, Bruchschollenland und Randfazies im Untergrund), in J. Lepper, and H.-G. Röhling, eds., Deutsche Stratigraphische Kommission: Stratigraphie von Deutschland XI. Buntsandstein: Schriftenreihe der Deutschen Gesellschaft für Geowissenschaften: Hannover, Schweizerbart, p. 547-582.

Füchtbauer, H., 1967, Der Einfluss des Ablagerungsmilieus auf die Sandsteindiagenese im Mittleren Buntsandstein: Sedimentary Geology, v. 1, p. 159-179.

Füchtbauer, H., 1974, Zur Diagenese fluviatiler Sandsteine: Geologische Rundschau, v. 63, p. 904-925.

Füchtbauer, H., 1988a, Diagenese, Sedimente und Sedimentgesteine: Sediment-Petrologie: Stuttgart, Schweizerbart, p. 147-184.

Füchtbauer, H., 1988b, Die Minerale, Sedimente und Sedimentgesteine: Sediment-Petrologie: Stuttgart, Schweizerbart, p. 97-128.

Füchtbauer, H., 1988c, Sedimente und Sedimentgesteine: Sediment-Petrologie: Stuttgart, Schweizerbart, 1141 p.

Garden, I. R., S. C. Guscott, S. D. Burley, K. A. Foxford, J. J. Walsh, and J. Marshall, 2001, An exhumed palaeo-hydrocarbon migration fairway in a faulted carrier system, Entrada Sandstone of SE Utah, USA: Geofluids, v. 1, p. 195-213.

Gaupp, R., A. Matter, J. Platt, K. Ramseyer, and J. Walzebuck, 1993, Diagenesis and fluid evolution of deeply buried Permian (Rotliegende) gas reservoirs, northwest Germany: AAPG Bulletin, v. 77, p. 1111-1128.

Gaupp, R., and J. A. Okkerman, 2011, Diagenesis and reservoir quality of Rotliegend sandstones in the northern Netherlands—A Review, in J. Grötsch, and R. Gaupp, eds., The Permian Rotliegend of The Netherlands: Tulsa, Oklahoma (USA), Society for Sedimentary Geology.

Gaupp, R., and R. Schöner, 2008, Intra-reservoir generation of organic acids and late stage enhanced porosity in sandstones, AAPG Annual Convention, San Antonio, Texas.

Geyer, G., 2002, Geologie von Unterfranken und angrenzenden Regionen: 5 Tabellen: Fränkische Landschaft: Gotha, Klett-Perthes, 588 p.

Geyer, G., 2005, The Fish River Subgroup in Namibia: stratigraphy, depositional environments and the Proterozoic–Cambrian boundary problem revisited: Geological Magazine, v. 142, p. 465-498.

Ghazi, S., and N. P. Mountney, 2009, Facies and architectural element analysis of a meandering fluvial succession: The Permian Warchha Sandstone, Salt Range, Pakistan: Sedimentary Geology, v. 221, p. 99-126.

Glaser, J., 1912, Geologische Untersuchungen im Gebiete der Emmendinger Vorberge. Südlicher Teil: Inaugural dissertation thesis, Albert-Ludwigs-Universität Freiburg, Freiburg i. Br, 1-146 p.

Grecula, M., S. S. Flint, H. D. V. Wickens, and S. D. Johnson, 2003, Upward-thickening patterns and lateral continuity of Permian sand-rich turbidite channel fills, Laingsburg Karoo, South Africa: Topographically controlled channel-overbank system: Sedimentology, v. 50, p. 831-853.

Greene, G. W., K. Kristiansen, E. E. Meyer, J. R. Boles, and J. N. Israelachvili, 2009, Role of electrochemical reactions in pressure solution: Geochimica et Cosmochimica Acta, v. 73, p. 2862-2874.

Günther, D., 2010, Der Schwarzwald und seine Umgebung: Geologie, Mineralogie, Bergbau ; Umwelt und Geotourismus: Sammlung geologischer Führer: Stuttgart, Borntraeger, 302 p.

Hagdorn, H., E. Nitsch, T. Aigner, and T. Simon, 2009, Field Guide: 6th International Triassic Field Workshop, p. 1-72.

Hartkamp-Bakker, C. A., and M. E. Donselaar, 2009, Permeability patterns in point bar deposits: Tertiary Loranca Basin, central Spain, in I. D. Bryant, and S. S. Flint, eds., The Geological Modelling of Hydrocarbon Reservoirs and Outcrop Analogues: Chichester, John Wiley & Sons, p. 157-168.

Haszeldine, R. S., O. Quinn, G. England, M. Wilkinson, Z. K. Shipton, J. P. Evans, J. Heath, L. Crossey, C. J. Ballentine, and C. M. Graham, 2005, Natural geochemical analogues for carbon dioxide storage in deep geological porous reservoirs, a United Kingdom perspective: Oil & Gas Science and Technology, v. 60, p. 33-49.

Hayes, M. J., and J. R. Boles, 1992, Volumetric relations between dissolved plagioclase and kaolinite in sandstones: Implications for aluminum mass transfer in the San Joaquin Basin, California, Origin, Diagenesis, and Petrophysics of Clay Minerals in Sandstone: Tulsa, Oklahoma (USA), Society of Sedimentary Geology, p. 111-123.

Heald, M. T., and R. E. Larese, 1974, Influence of coatings on quartz cementation: Journal of Sedimentary Petrology, v. 44, p. 1269-1274.

Heggemann, H., K.-W. Tietze, and H.-J. Behr, 2003, The river system of the Phra Wihan Formation, Thailand: Göttinger Arbeiten zur Geologie und Paläontologie, v. 5, p. 23-32.

Henares, S., J. Arribas, G. Cultrone, and C. Viseras, 2016a, Muddy and dolomitic rip-up clasts in Triassic fluvial sandstones: Origin and impact on potential reservoir properties (Argana Basin, Morocco): Sedimentary Geology, v. 339, p. 218-233.

Henares, S., L. Caracciolo, C. Viseras, J. Fernández, and L. M. Yeste, 2016b, Diagenetic constraints on heterogeneous reservoir quality assessment: A Triassic outcrop analog of meandering fluvial reservoirs: AAPG Bulletin, v. 100, p. 1377-1398.

Hilse, U., A. Goepel, D. Pudlo, K. Heide, and R. Gaupp, 2010, Characterization of CO2-induced (?) bleaching phenomena in German red bed sediments by combined geochemical and evolved gas analysis: EGU General Assembly 2010, Vienna, p. 1684.

Hofmann, R., and A. Baumann, 1984, Preliminary report on the Sr isotopic composition of hydrothermal vein barites in the Federal Republic of Germany: Mineralium Deposita, v. 19, p. 166-169.

Hornung, J., and T. Aigner, 1999, Reservoir and aquifer characterization of fluvial architectural elements: Stubensandstein, Upper Triassic, southwest Germany: Sedimentary Geology, v. 129, p. 215-280.

Hornung, J., and T. Aigner, 2004, Sedimentäre Architektur und Poroperm-Analyse fluviatiler Sandsteine: Fallbeispiel Coburger Sandstein, Franken: Hallesches Jahrbuch für Geowissenschaften, v. 18, p. 121-138.

Howarth, R. J., 1998, Improved estimators of uncertainty in proportions, point-counting, and pass-fail test results: American Journal of Science, v. 298, p. 594-607.

Hsü, K. J., 1977, Studies of Ventura Field, California, II: Lithology, compaction, and permeability of sands: AAPG Bulletin, v. 61.

Huang, W. L., A. M. Bishop, and R. W. Brown, 1986, The effect of fluid/rock ratio on feldspar dissolution and illite formation under reservoir conditions: Clay Minerals, v. 21, p. 585-601.

Hurst, A., and P. H. Nadeau, 1995, Clay microporosity in reservoir sandstones: An application of quantitative electron microscopy in petrophysical evaluation: AAPG Bulletin, v. 79, p. 563-573.

Kasch, N., J. Kley, J. Köster, and J. Wendler, 2010, Fracture-related fluid migration and fluid-rock interaction in outcrop analogues of Buntsandstein reservoir rocks (southern Thuringia and northern Hesse): EGU General Assembly 2010, Vienna, p. 13194.

Keßler, G., and J. Leiber, 1994, Erläuterungen zu Blatt 7613 Lahr/Schw.-Ost: Geologische Karte 1 : 25000 von Baden-Württemberg: Freiburg, Geologisches Landesamt Baden-Württemberg.

Kirchner, H., 1927, Über die Tierfährten im oberen Buntsandstein: Paläontologische Zeitschrift, v. 9, p. 112-121.

Kirkland, D. W., R. E. Denison, and M. A. Rooney, 1995, Diagenetic alteration of Permian strata at oil fields of south central Oklahoma, USA: Marine and Petroleum Geology, v. 12, p. 629-644.

Klingbeil, R., S. Kleineidam, U. Asprion, T. Aigner, and G. Teutsch, 1999, Relating lithofacies to hydrofacies: outcrop-based hydrogeological characterisation of Quaternary gravel deposits: Sedimentary Geology, v. 129, p. 299-310.

Klinkenberg, L. J., 1941, The permeability of porous media to liquids and gases, Drilling and Production Practice, New York, American Petroleum Institute, p. 200-213.

Komatsubara, J., 2004, Fluvial architecture and sequence stratigraphy of the Eocene to Oligocene Iwaki Formation, northeast Japan: channel-fills related to the sea-level change: Sedimentary Geology, v. 168, p. 109-123.

Kristiansen, K., M. Valtiner, G. W. Greene, J. R. Boles, and J. N. Israelachvili, 2011, Pressure solution – The importance of the electrochemical surface potentials: Geochimica et Cosmochimica Acta, v. 75, p. 6882-6892.

Kunkel, C., M. Aehnelt, D. Pudlo, N. Kukowski, K. U. Totsche, and R. Gaupp, 2018, Subsurface aquifer heterogeneities of Lower Triassic clastic sediments in central Germany: Marine and Petroleum Geology, v. 97, p. 209-222.

Lander, R. H., and L. M. Bonnell, 2010, A model for fibrous illite nucleation and growth in sandstones: AAPG Bulletin, v. 94, p. 1161-1187.

Lander, R. H., R. E. Larese, and L. M. Bonnell, 2008, Toward more accurate quartz cement models: The importance of euhedral versus noneuhedral growth rates: AAPG Bulletin, v. 92, p. 1537-1563.

Lander, R. H., and O. Walderhaug, 1999, Predicting porosity through simulating sandstone compaction and quartz cementation: AAPG Bulletin, v. 83, p. 433-449.

Lanson, B., D. Beaufort, G. Berger, A. Bauer, A. Cassagnabère, and A. Meunier, 2002, Authigenic kaolin and illitic minerals during burial diagenesis of sandstones: a review: Clay Minerals, v. 37, p. 1-22.

Le Blanc Smith, G., 1980, Logical-letter coding system for facies nomenclature: Witbank coalfield: Transactions of the Geological Society of South Africa, v. 83, p. 301-311.

Leiber, J., and H. Bock, 2013, Der Buntsandstein in der Kraichgau-Senke und der südwestdeutschen Randfazies (Baden-Württemberg), in J. Lepper, and H.-G. Röhling, eds., Deutsche Stratigraphische Kommission: Stratigraphie von Deutschland XI. Buntsandstein: Schriftenreihe der Deutschen Gesellschaft für Geowissenschaften: Hannover, Schweizerbart, p. 525-546.

Liewig, N., and N. Clauer, 2000, K-Ar dating of varied microtextural illite in Permian gas reservoirs, northern Germany: Clay Minerals, v. 35, p. 271-281.

Line, L. H., J. Jahren, and H. Hellevang, 2018, Mechanical compaction in chlorite-coated sandstone reservoirs – Examples from Middle – Late Triassic channels in the southwestern Barents Sea: Marine and Petroleum Geology, v. 96, p. 348-370.

Lippmann, R., 2012, Diagenesis in Rotliegend, Triassic and Jurassic clastic hydrocarbon reservoirs of the Central Graben, North Sea: Doctoral thesis, Friedrich-Schiller-Universität, Jena, 283 p.

Lovley, D. R., E. E. Roden, E. J. P. Philips, and J. C. Woodward, 1993, Enzymatic iron and uranium reduction by sulfate-reducing bacteria: Marine Geology, v. 113, p. 41-53.

Lü, Z.-X., S.-J. Ye, X. Yang, R. Li, and Y.-H. Qing, 2015, Quantification and timing of porosity evolution in tight sand gas reservoirs: an example from the Middle Jurassic Shaximiao Formation, western Sichuan, China: Petroleum Science, v. 12, p. 207-217.

Lundegard, P. D., 1992, Sandstone porosity loss - a "big picture" view of the importance of compaction: Journal of Sedimentary Petrology, v. 62, p. 250-260.

Machel, H. G., 2001, Bacterial and thermochemical sulfate reduction in diagenetic settings - old and new insights: Sedimentary Geology, v. 140, p. 143-175.

Mader, D., and T. Teyssen, 1985, Palaeoenvironmental interpretation of fluvial red beds by statistical analysis of paleocurrent data: Examples from the Buntsandstein (Lower Triassic) of the Eifel and Bavaria in the German Basin (Middle Europe): Sedimentary Geology, v. 41, p. 1-74.

Makowitz, A., R. H. Lander, and K. L. Milliken, 2006, Diagenetic modeling to assess the relative timing of quartz cementation and brittle grain processes during compaction: AAPG Bulletin, v. 90, p. 873-885.

Makowitz, A., and D. Sibley, 2001, Crystal growth mechanisms of quartz overgrowths in a Cambrian quartz arenite: Journal of Sedimentary Research, v. 71, p. 809-816.

Martini, E., 1992, Fossilien, Sedimentmarken und Palökologie im Oberen Buntsandstein (Röt) der Rhön: Natur und Museum, v. 122, p. 90-99.

Marzo, M., W. Nijman, and C. Puigdefabregas, 1988, Architecture of the Castissent fluvial sheet sandstones, Eocene, South Pyrenees, Spain: Sedimentology, v. 35, p. 719-738.

Mast, R. F., and P. E. Potter, 1963, Sedimentary structures, sand shape fabrics, and permeability. II: The Journal of Geology, v. 71, p. 548-565.

McBride, E. F., 1963, A classification of common sandstones: Journal of Sedimentary Petrology, v. 33, p. 664-669.

McKinley, J. M., P. M. Atkinson, C. D. Lloyd, A. H. Ruffell, and R. H. Worden, 2011, How porosity and permeability vary spatially with grain size, sorting, cement volume, and mineral dissolution in fluvial Triassic sandstones: The value of geostatistics and local regression: Journal of Sedimentary Research, v. 81, p. 844-858.

Menning, M., 2018, Die Stratigraphische Tabelle von Deutschland 2016 (STD 2016) / The Stratigraphic Table of Germany 2016 (STG 2016): Zeitschrift der Deutschen Gesellschaft für Geowissenschaften, v. 169, p. 105-128.

Meschede, M., and L. N. Warr, 2019, The Geology of Germany - A Process-Oriented Approach: Regional Geology Reviews: Cham (Switzerland), Springer Nature.

Meyer, R., and F. F. Krause, 2006, Permeability anisotropy and heterogeneity of a sandstone reservoir analogue: An estuarine to shoreface depositional system in the Virgelle Member, Milk River Formation, Writing-on-Stone Provincial Park, southern Alberta: Bulletin of Canadian Petroleum Geology, v. 54, p. 301-318.

Miall, A. D., 1977a, Lithofacies types and vertical profile models in braided river deposits: A summary, in A. D. Miall, ed., Fluvial Sedimentology: Memoir - Canadian Society of Petroleum Geologists, Stacs Data Service Ltd, p. 597-604.

Miall, A. D., 1977b, A review of the braided-river depositional environment: Earth-Science Reviews, v. 13, p. 1-62.

Miall, A. D., 1985, Architectural-element analysis: A new method of facies analysis applied to fluvial deposits: Earth-Science Reviews, v. 22, p. 261-308.

Miall, A. D., 1988a, Architectural elements and bounding surfaces in fluvial deposits: Anatomy of the Kayenta Formation (Lower Jurassic), Southwest Colorado: Sedimentary Geology, v. 55, p. 233-262.

Miall, A. D., 1988b, Reservoir heterogeneities in fluvial sandstones: Lessons from outcrop studies: AAPG Bulletin, v. 72, p. 682-697.

Miall, A. D., 1996, The Geology of Fluvial Deposits: Sedimentary Facies, Basin Analysis, and Petroleum Geology: Berlin, New York, Springer, 582 p.

Miall, A. D., 2014, The facies and architecture of fluvial systems, in A. D. Miall, ed., Fluvial Depositional Systems: Cham (Switzerland), Springer International Publishing, p. 25-35.

Miall, A. D., and C. E. Turner-Peterson, 1989, Variations in fluvial style in the Westwater Canyon Member, Morrison Formation (Jurassic), San Juan Basin, Colorado Plateau: Sedimentary Geology, v. 63, p. 21-60.

Molenaar, N., 1984, Palaeopedogenic features and their palaeoclimatological significance for the Nevremont Formation (Lower Givetian), the Northern Ardennes, Belgium: Palaeogeography, Palaeoclimatology, Palaeoecology, v. 46, p. 325-344.

Molenaar, N., 1986, The interrelation between clay infiltration, quartz cementation, and compaction in Lower Givetian terrestrial sandstones, northern Ardennes, Belgium: Journal of Sedimentary Research, v. Vol. 56.

Molenaar, N., and M. Felder, 2018, Clay cutans and the origin of illite rim cement: An example from the siliciclastic Rotliegend sandstone in the Dutch Southern Permian Basin: Journal of Sedimentary Research, v. 88, p. 641-658.

Molenaar, N., M. Felder, K. Bär, and A. E. Götz, 2015, What classic greywacke (litharenite) can reveal about feldspar diagenesis: An example from Permian Rotliegend sandstone in Hessen, Germany: Sedimentary Geology, v. 326, p. 79-93.

Monsees, A. C., B. Busch, N. Schöner, and C. Hilgers, 2020, Rock typing of diagenetically induced heterogeneities – A case study from a deeply-buried clastic Rotliegend reservoir of the Northern German Basin: Marine and Petroleum Geology, v. 113, p. 104163.

Morad, S., K. Al-Ramadan, J. M. Ketzer, and L. F. De Ros, 2010, The impact of diagenesis on the heterogeneity of sandstone reservoirs: A review of the role of depositional facies and sequence stratigraphy: AAPG Bulletin, v. 94, p. 1267-1309.

Morad, S., and A. A. Aldahan, 1986, Alteration of detrital Fe-Ti oxides in sedimentary rocks: Geological Society of America Bulletin, v. 97, p. 567-578.

Morad, S., and A. A. Aldahan, 1987, Diagenetic "replacement" of feldspars by titanium oxides in sandstones: Sedimentary Geology, v. 51, p. 147-153.

Morad, S., R. Marfil, and J. A. de la Peña, 1989, Diagenetic K-feldspar pseudomorphs in the Triassic Buntsandstein sandstones of the Iberian Range, Spain: Sedimentology, v. 36, p. 635-650.

Moraes, M. A. S., and R. C. Surdam, 1993, Diagenetic heterogeneity and reservoir quality: Fluvial, deltaic, and turbiditic sandstone reservoirs, Potiguar and Reconcavo Rift Basins, Brazil: AAPG Bulletin, v. 77, p. 1142-1158.

North, C. P., and M. Boering, 1999, Spectral gamma-ray logging for facies discrimination in mixed fluvial-eolian successions: A cautionary tale: AAPG Bulletin, v. 83, p. 155-169.

Nover, G., J. Von Der Gönna, S. Heikamp, and J. Köster, 2013, Changes of petrophysical properties of sandstones due to interaction with supercritical carbon dioxide – a laboratory study: European Journal of Mineralogy, v. 25, p. 317-329.

Okrusch, M., G. Geyer, J. Lorenz, J. Jung, and S. Matthes, 2011, Spessart: geologische Entwicklung und Struktur, Gesteine und Minerale: Sammlung geologischer Führer, v. Bd. 106: Stuttgart, Borntraeger, 368 p.

Ortlam, D., 1974, Inhalt und Bedeutung fossiler Bodenkomplexe in Perm und Trias von Mitteleuropa: Geologische Rundschau, v. 63, p. 850-884.

Parry, W. T., M. A. Chan, and B. Beitler, 2004, Chemical bleaching indicates episodes of fluid flow in deformation bands in sandstone: AAPG Bulletin, v. 88, p. 175-191.

Paul, J., 1982, Der Untere Buntsandstein des Germanischen Beckens: Geologische Rundschau, v. 71, p. 795-811.

Paxton, S. T., J. O. Szabo, J. M. Ajdukiewicz, and R. E. Klimentidis, 2002, Construction of an intergranular volume compaction curve for evaluating and predicting compaction and porosity loss in rigid-grain sandstone reservoirs: AAPG Bulletin, v. 86, p. 2047-2067.

Pearce, J. M., G. A. Kirby, A. Lacinska, L. Bateson, D. Wagner, C. A. Rochelle, and M. Cassidy, 2011, Reservoir-scale CO2 -fluid rock interactions: Preliminary results from field investigations in the Paradox Basin, Southeast Utah: Energy Procedia, v. 4, p. 5058-5065.

Peschel, S. E., 2013, Der Einfluss des Forster Vulkanismus auf die Diagenese des gebleichten Buntsandsteins des Pfälzer Waldes parallel zum Rheingrabenrand: Diploma thesis, Friedrich-Schiller-Universität, Jena, 129 p.

Pranter, M. J., A. I. Ellison, R. D. Cole, and P. E. Patterson, 2007, Analysis and modeling of intermediate-scale reservoir heterogeneity based on a fluvial point-bar outcrop analog, Williams Fork Formation, Piceance Basin, Colorado: AAPG Bulletin, v. 91, p. 1025-1051.

Pudlo, D., V. Reitenbach, D. Albrecht, L. Ganzer, U. Gernert, J. Wienand, B. Kohlhepp, and R. Gaupp, 2012, The impact of diagenetic fluid–rock reactions on Rotliegend sandstone composition and petrophysical properties (Altmark area, central Germany): Environmental Earth Sciences, v. 67, p. 369-384.

Regierungspräsidium Freiburg - Landesamt für Geologie Rohstoffe und Bergbau, 2019, LGRB-Kartenviewer.

Reinecker, J., R. Grobe, T. Hochschild, J. Bauer, S. Meier, S. Philipp, M. Filomena, H. Stollhofen, T. Bechstädt, T. Drews, G. Miernik, D. Soyk, and B. Melchert, 2015, Verbundprojekt AuGE: Aufschlussanalogstudien und ihre Anwendbarkeit in der geothermischen Exploration - Entwicklung von Methoden zur Ermittlung von Permeabilitäten und Transmissivitäten aus Reservoir-Informationen des Oberrheingrabens, Karlsruhe (Germany), GeoThermal Engineering GmbH, p. 1-75.

Rider, M. H., and M. Kennedy, 2011, The Geological Interpretation of Well Logs: Sutherland (Scotland), Rider-French Consulting Ltd., 432 p.

Rieckmann, M., 1970, Untersuchung von Turbulenzerscheinungen beim Fließen von Gasen durch Speichergesteine unter Berücksichtigung der Gleitströmung: Erdoel-Erdgas-Zeitschrift, v. 86, p. 36-51.

Rimstidt, J. D., and H. L. Barnes, 1980, The kinetics of silica-water reactions: Geochimica et Cosmochimica Acta, v. 44, p. 1683-1699.

Röhling, H.-G., and J. Lepper, 2013, Paläogeographie des Mitteleuropäischen Beckens während der tieferen Trias (Buntsandstein), in J. Lepper, and H.-G. Röhling, eds., Deutsche Stratigraphische Kommission: Stratigraphie von Deutschland XI: Buntsandstein: Schriftenreihe der Deutschen Gesellschaft für Geowissenschaften: Hannover, Schweizerbart, p. 43-67.

Röhling, H.-G., J. Lepper, M. Diehl, D. Dittrich, W. Freudenberger, V. Friedlein, N. Hug-Diegel, and E. Nitsch, 2018, Der Buntsandstein in der Stratigraphischen Tabelle von Deutschland 2016: Zeitschrift der Deutschen Gesellschaft für Geowissenschaften, v. 169, p. 151-180.

Salem, A. M., S. Morad, L. F. Mato, and I. S. Al-Aasm, 2000, Diagenesis and reservoir-quality evolution of fluvial sandstones during progressive burial and uplift: Evidence from the Upper Jurassic Boipeba Member, Reconcavo Basin, northeastern Brazil: AAPG Bulletin, v. 84.

Sambrook Smith, G. H., J. L. Best, J. Z. Leroy, and O. Orfeo, 2016, The alluvial architecture of a suspended sediment dominated meandering river: The Río Bermejo, Argentina: Sedimentology, v. 63, p. 1187-1208.

Scherer, M., 1987, Parameters influencing porosity in sandstones: A model for sandstone porosity prediction: AAPG Bulletin, v. 71, p. 485-491.

Schmidt, C., B. Busch, and C. Hilgers, 2020a, Compaction and cementation control on bleaching in Triassic fluvial red beds, S-Germany: Zeitschrift der Deutschen Gesellschaft für Geowissenschaften, p. 93573.

Schmidt, C., B. Busch, and C. Hilgers, 2020b, Lateral variations of detrital, authigenic and petrophysical properties in an outcrop analog of the fluvial Plattensandstein, Lower Triassic, Central S-Germany: Zeitschrift der Deutschen Gesellschaft für Geowissenschaften, p. 95382.

Schöner, R., and R. Gaupp, 2005, Contrasting red bed diagenesis: the southern and northern margin of the Central European Basin: International Journal of Earth Sciences, v. 94, p. 897-916.

Schreiber, B. C., and M. El Tabakh, 2000, Deposition and early alteration of evaporites: Sedimentology, v. 47, p. 215-238.

Schröder, B., 1982, Entwicklung des Sedimentbeckens und Stratigraphie der klassischen Germanischen Trias: Geologische Rundschau, v. 71, p. 783-794.

Schultz-Krutisch, T., and F. Heller, 1985, Measurement of magnetic susceptibility anisotropy in Buntsandstein deposits from southern Germany: Journal of Geophysics, v. 56, p. 51-58.

Schumacher, D., 1996, Hydrocarbon-induced alteration of soils and sediments, in D. Schumacher, and M. A. Abrams, eds., Hydrocarbon Migration and Its Near-Surface Expression: Outgrowth of the AAPG Hedberg Research Conference, Vancouver, British Columbia, April 24-28, 1994: AAPG Memoir, v. 66, p. 71-89.

Shebl, M. A., and R. C. Surdam, 1996, Redox reactions in hydrocarbon clastic reservoirs: experimental validation of this mechanism for porosity enhancement: Chemical Geology, v. 132, p. 103-117.

Sheldon, H. A., J. Wheeler, R. H. Worden, and M. J. Cheadle, 2003, An analysis of the roles of stress, temperature, and pH in chemical compaction of sandstones: Journal of Sedimentary Research, v. 73, p. 64-71.

Sindern, S., V. Havenith, A. Gerdes, F. M. Meyer, D. Adelmann, and A. Hellmann, 2019, Dating of anatase-forming diagenetic reactions in Rotliegend sandstones of the North German Basin: International Journal of Earth Sciences, v. 108, p. 1275-1292.

Staude, S., P. D. Bons, and G. Markl, 2009, Hydrothermal vein formation by extension-driven dewatering of the middle crust: An example from SW Germany: Earth and Planetary Science Letters, v. 286, p. 387-395.

Staude, S., S. Göb, K. Pfaff, F. Ströbele, W. R. Premo, and G. Markl, 2011, Deciphering fluid sources of hydrothermal systems: A combined Sr- and S-isotope study on barite (Schwarzwald, SW Germany): Chemical Geology, v. 286, p. 1-20.

Stewart, D. J., 1981, A meander-belt sandstone of the Lower Cretaceous of Southern England: Sedimentology, v. 28, p. 1-20.

Tanner, W. F., 1967, Ripple mark indices and their uses: Sedimentology, v. 9, p. 89-104.

Taylor, A. W., and B. D. Ritts, 2004, Mesoscale heterogeneity of fluvial-lacustrine reservoir analogues: Examples from the Eocene Green River and Colton Formations, Uinta Basin, Utah, USA: Journal of Petroleum Geology, v. 27, p. 3-26.

Taylor, J. M., 1950, Pore-space reduction in sandstones: AAPG Bulletin, v. 34, p. 701-716.

Taylor, S., and R. Barker, 2002, Resistivity of partially saturated Triassic Sandstone: Geophysical Prospecting, v. 50, p. 603-613.

Taylor, T., R. Stancliffe, C. Macaulay, and L. Hathon, 2004, High temperature quartz cementation and the timing of hydrocarbon accumulation in the Jurassic Norphlet sandstone, offshore Gulf of Mexico, USA: Geological Society, London, Special Publications, v. 237, p. 257-278.

Taylor, T. R., M. R. Giles, L. A. Hathon, T. N. Diggs, N. R. Braunsdorf, G. V. Birbiglia, M. G. Kittridge, C. I. Macaulay, and I. S. Espejo, 2010, Sandstone diagenesis and reservoir quality prediction: Models, myths, and reality: AAPG Bulletin, v. 94, p. 1093-1132.

Thyne, G., B. P. Boudreau, M. Ramm, and R. E. Midtbø, 2001, Simulation of potassium feldspar dissolution and illitization in the Statfjord Formation, North Sea: AAPG Bulletin, v. 85, p. 621-635.

Tietze, K.-W., 1982, Zur Geometrie einiger Flüsse im Mittleren Buntsandstein (Trias): Geologische Rundschau, v. 71, p. 813-828.

Trendell, A. M., S. C. Atchley, and L. C. Nordt, 2012, Depositional and diagenetic controls on reservoir attributes within a fluvial outcrop analog: Upper Triassic Sonsela member of the Chinle Formation, Petrified Forest National Park, Arizona: AAPG Bulletin, v. 96, p. 679-707.

Trusheim, F., 1937, Wurzelböden im Plattensandstein Mainfrankens: Zentralblatt für Mineralogie, Geologie und Paläontologie, Abt. B, v. 10, p. 385-388.

Tucker, M. E., 2003, Sedimentary rocks in the field: The geological field guide series: West Sussex, England; Hoboken, NJ, J. Wiley, 234 p.

Tyler, N., and R. J. Finley, 1991, Architectural controls on the recovery of hydrocarbons from sandstone reservoirs, The three-dimensional facies architecture of terrigenous clastic sediments, and its implications for hydrocarbon discovery and recovery: Concepts in Sedimentology and Paleontology, v. 3, Society of Economic Paleontologists and Mineralogists, p. 1-5.

Udden, J. A., 1914, Mechanical composition of clastic sediments: Bulletin of the Geological Society of America, v. 25, p. 655-744.

Van den Berg, J. H., and A. Van Gelder, 1993, A new bedform stability diagram, with emphasis on the transition of ripples to plane bed in flows over fine sand and silt, in M. Marzo, and C. Puigdefábregas, eds., Alluvial Sedimentation: Special Publicatons of the International Association of Sedimentologists: Oxford (England); Boston (USA), Blackwell Scientific Publications, p. 11-21.

Veiga, G. D., L. A. Spalletti, and S. Flint, 2002, Aeolian/fluvial interactions and high-resolution sequence stratigraphy of a non-marine lowstand wedge: the Avilé Member of the Agrio Formation (Lower Cretaceous), central Neuquén Basin, Argentina: Sedimentology, v. 49, p. 1001-1019.

Wachutka, M., and T. Aigner, 2001, Reservoirgeologie und Petrophysik: Eine Aufschluss-Analogstudie im Buntsandstein (Trias, Schwarzwald, SW-Deutschland): Zeitschrift der angewandten Geologie, v. 47, p. 191-198.

Walderhaug, O., 1996, Kinetic modeling of quartz cementation and porosity loss in deeply buried sandstone reservoirs: AAPG Bulletin, v. 80, p. 731-745.

Walderhaug, O., 2000, Modeling quartz cementation and porosity in Middle Jurassic Brent Group sandstones of the Kvitebjørn Field, northern North Sea: AAPG Bulletin, v. 84, p. 1325-1339.

Walderhaug, O., and P. A. Bjorkum, 2003, The effect of stylolite spacing on quartz cementation in the Lower Jurassic Sto Formation, southern Barents Sea: Journal of Sedimentary Research, v. 73, p. 146-156.

Walderhaug, O., A. Eliassen, and N. E. Aase, 2012, Prediction of permeability in quartz-rich sandstones: Examples from the Norwegian continental shelf and the Fontainebleau Sandstone: Journal of Sedimentary Research, v. 82, p. 899-912.

Walker, T. R., 1967a, Color of recent sediments in tropical Mexico: A contribution to the origin of red beds: Geological Society of America Bulletin, v. 78, p. 917-920.

Walker, T. R., 1967b, Formation of red beds in modern and ancient deserts: Geological Society of America Bulletin, v. 78, p. 353-368.

Walker, T. R., 1975, Red beds in the western interior of the United States, in E. D. McKee, and E. Crosby, J., eds., Paleotectonic Investigations of the Pennsylvanian System in the United States: Geological Survey Professional Paper: Washington, United States Government Printing Office, p. 49-56.

Walker, T. R., B. Waugh, and A. J. Grone, 1978, Diagenesis in first-cycle desert alluvium of Cenozoic age, southwestern United States and northwestern Mexico: Geological Society of America Bulletin, v. 89, p. 19-32.

Wang, D., Y. Song, and M. Zhao, 2012, Influence of the volume loss of framework grains on the quantitative analysis of diagenetic modification of the original intergranular porosity: Marine and Petroleum Geology, v. 30, p. 26-31.

Weber, J., and J. Lepper, 2002, Depositional environment and diagenesis as controlling factors for petro-physical properties and weathering resistance of siliciclastic dimension stones: integrative case study on the 'Wesersandstein' (northern Germany, Middle Buntsandstein), in S. Siegesmund, T. N. Weiss, and A. Vollbrecht, eds., Natural Stone, Weathering Phenomena, Conservation Strategies and Case Studies: Geological Society special publication: London, The Geological Society of London, p. 103-114.

Weibel, R., 1998, Diagenesis in oxidising and locally reducing conditions - an example from the Triassic Skagerrak Formation, Denmark: Sedimentary Geology, v. 121, p. 259-276.

Weibel, R., 1999, Effects of burial on the clay assemblages in the Triassic Skagerrak Formation, Denmark: Clay Minerals, v. 34, p. 619-635.

Wendler, J., J. Köster, J. Götze, N. Kasch, N. Zisser, J. Kley, D. Pudlo, G. Nover, and R. Gaupp, 2012, Carbonate diagenesis and feldspar alteration in fracture-related bleaching zones (Buntsandstein, central Germany): possible link to CO2-influenced fluid–mineral reactions: International Journal of Earth Sciences, v. 101, p. 159-176.

Wentworth, C. K., 1922, A scale of grade and class terms for clastic sediments: The Journal of Geology, v. 30, p. 377-392.

Werner, W., H. Bock, J. Wittenbrink, and B. Kimmig, 2013, Buntsandstein, in R. u. B. Baden-Württemberg. Landesamt f. Geologie, ed., Naturwerksteine aus Baden-Württemberg: Vorkommen, Beschaffenheit und Nutzung; mit 45 Tabellen: Freiburg i. Br., Landesamt für Geologie, Rohstoffe und Bergbau, p. 765.

Werner, W., and V. Dennert, 2004, Lagerstätten und Bergbau im Schwarzwald: ein Führer unter besonderer Berücksichtigung der für die Öffentlichkeit zugänglichen Bergwerke: Freiburg i. Br., Landesamt für Geologie, Rohstoffe und Bergbau Baden-Württemberg, 334 p.

Werner, W., and H. J. Franzke, 2001, Postvariszische bis neogene Bruchtektonik und Mineralisation im südlichen Zentralschwarzwald: Zeitschrift der Deutschen Geologischen Gesellschaft, v. 152, p. 405-437.

Werner, W., H. J. Franzke, G. Wirsing, J. Jochum, V. Lüders, and J. Wittenbrink, 2002, Die Erzlagerstätte Schauinsland bei Freiburg im Breisgau - Bergbau, Geologie, Hydrogeologie, Mineralogie, Geochemie, Tektonik und Lagerstättenentstehung: Berichte der Naturforschenden Gesellschaft zu Freiburg i. Br., v. 92, p. 75-83.

Wilkinson, M., R. S. Haszeldine, and A. E. Fallick, 2006, Jurassic and Cretaceous clays of the northern and central North Sea hydrocarbon reservoirs reviewed: Clay Minerals, v. 41, p. 151-186.

Wilson, A. M., M. Huettel, and S. Klein, 2008, Grain size and depositional environment as predictors of permeability in coastal marine sands: Estuarine, Coastal and Shelf Science, v. 80, p. 193-199.

Wilson, J. C., and E. F. McBride, 1988, Compaction and porosity evolution of Pliocene sandstones, Ventura Basin, California: AAPG Bulletin, v. 72, p. 664-681.

Wooldridge, L. J., R. H. Worden, J. Griffiths, and J. E. P. Utley, 2017, Clay-coated sand grains in petroleum reservoirs: Understanding their distribution via a modern analogue: Journal of Sedimentary Research, v. 87, p. 338-352.

Wooldridge, L. J., R. H. Worden, J. Griffiths, and J. E. P. Utley, 2019, How to quantify clay-coat grain coverage in modern and ancient sediments: Journal of Sedimentary Research, v. 89, p. 135-146.

Worden, R. H., P. J. Armitage, and A. R. Butcher, 2018, Petroleum reservoir quality prediction: overview and contrasting approaches from sandstone and carbonate communities, in J. M. Churchill, A. E. Csoma, C. Hollis, R. H. Lander, and J. E. Omma, eds., Reservoir Quality of Clastic and Carbonate Rocks: Analysis, Modelling and Prediction: Geological Society special publication: London, The Geological Society, p. 1-31.

Worden, R. H., and S. D. Burley, 2003, Sandstone diagenesis: The evolution of sand to stone, in S. D. Burley, and R. H. Worden, eds., Sandstone Diagenesis: Recent and Ancient: Reprint series volume 4 of the International Association of Sedimentologists: Malden, Massachusetts (USA), Blackwell Pub, p. 3-44.

Worden, R. H., and S. Morad, 2003, Clay minerals in sandstones: controls on formation, distribution and evolution, in R. H. Worden, and S. Morad, eds., Clay Mineral Cements in Sandstones: Special publication number 34 of the International Association of Sedimentologists: Malden, Massachusetts (USA), Blackwell Pub, p. 520.

Wüstefeld, P., U. Hilse, B. Koehrer, D. Adelmann, and C. Hilgers, 2017, Critical evaluation of an Upper Carboniferous tight gas sandstone reservoir analog: Diagenesis and petrophysical aspects: Marine and Petroleum Geology, v. 86, p. 689-710.

Ziegler, K., 2006, Clay minerals of the Permian Rotliegend Group in the North Sea and adjacent areas: Clay Minerals, v. 41, p. 355-393.

Ziegler, P. A., 1982, Triassic rifts and facies patterns in Western and Central Europe: Geologische Rundschau, v. 71, p. 747-772.

Ziegler, P. A., 1990, Geological Atlas of Western and Central Europe: The Hague, Shell Internationale Petroleum Maatschappij B.V., 239 p.

8 Appendix

Table A.1: Compilation of point count, petrophysical data and additional sample information (Chapter 2, 3).

Sample Name	Lithostratigraphic unit	Facies	Sample type	Bleaching type	*Point counts*	*Detrital components [%]*	Quartz	Alkali feldspar	Shale RF	Siltstone RF	Sandstone RF	Chert undifferentiated	Silicic VRF	altered Volcanic RF	Phyllite	Metamorphic RF undifferentiated	Plutonic RF undifferentiated	Lithoclast undifferentiated	Mica	Sericite
RB-01	soPl	Fluvial	MB	cu			36.3	6.0	0.7	0.0	0.0	0.7	0.0	0.7	1.3	0.0	0.7	0.3	1.3	0.3
RB-02	soPl	Fluvial	MB	cu			38.7	8.7	1.0	0.0	0.0	0.0	0.7	0.3	2.0	0.0	0.3	0.0	2.3	0.0
RB-03	soPl	Fluvial	MB	cu			37.3	6.3	1.0	0.0	0.0	0.7	0.3	0.0	2.0	0.0	0.0	0.0	1.3	0.3
RB-04	soPl	Fluvial	MB	cu			37.5	11.7	1.3	1.0	0.8	0.3	1.0	0.0	1.2	1.0	0.0	0.0	0.8	0.3
RB-05	soPl	Fluvial	MB	cu			35.3	11.7	0.3	0.0	0.0	0.7	0.3	0.0	2.0	1.7	0.0	0.0	0.7	0.0
RB-06	soPl	Fluvial	MB	fb			38.0	9.3	0.3	0.3	0.0	0.3	1.3	0.0	2.0	1.7	0.0	0.0	1.7	0.0
RB-07	soPl	Fluvial	MB	fb			38.3	9.0	1.0	0.3	0.0	0.3	1.0	0.0	1.7	0.3	0.0	0.0	1.0	0.0
RB-08	soPl	Fluvial	QW	cu			34.3	8.3	0.0	0.0	0.0	0.0	0.7	0.7	2.3	1.0	0.0	0.0	1.3	0.3
RB-09	soPl	Fluvial	QW	cu			37.7	9.3	0.0	0.0	0.0	0.0	0.7	0.0	0.7	2.3	0.0	0.0	0.7	0.7
RB-10	soPl	Fluvial	QW	cu			32.3	6.0	0.7	0.0	0.0	0.3	0.0	0.0	1.0	2.7	0.0	0.0	1.3	0.3
RB-11	soPl	Fluvial	QW	cu			32.7	9.7	0.3	0.0	0.0	0.0	0.0	0.0	0.3	3.0	0.0	0.0	1.0	0.3
RB-12	soPl	Fluvial	QW	cu			33.0	9.7	0.3	0.0	0.0	0.0	0.0	0.0	1.3	3.0	0.0	0.0	1.0	0.7
RB-13	soPl	Fluvial	QW	cu			29.3	7.0	0.3	0.3	0.0	0.0	2.0	0.0	0.7	1.7	0.0	0.0	3.3	0.0
RB-14	soPl	Fluvial	QW	cu			36.0	7.3	0.7	0.0	0.3	0.0	0.3	0.0	0.0	1.3	0.0	0.0	2.0	0.7
RB-15	soPl	Fluvial	QW	cu			37.3	6.7	1.7	0.0	0.0	0.7	1.0	0.0	0.7	2.0	0.0	0.0	0.7	0.3
RB-16	soPl	Fluvial	QW	cu			41.0	6.3	0.3	0.0	0.0	0.0	1.0	0.0	0.7	1.3	0.0	0.0	0.7	0.0
RB-17	soPl	Fluvial	QW	pb			N/A	N/A	N/A	N/A	N/A	N/A	N/A	N/A	N/A	N/A	N/A	N/A	N/A	N/A
RB-18	soPl	Fluvial	QW	cu			28.3	8.7	0.3	1.0	0.0	0.0	1.0	0.0	1.3	1.3	0.3	0.0	2.7	0.0
RB-19	soPl	Fluvial	QW	cu			N/A	N/A	N/A	N/A	N/A	N/A	N/A	N/A	N/A	N/A	N/A	N/A	N/A	N/A
RB-20	soPl	Fluvial	QW	cu			32.0	10.7	0.0	0.3	0.0	0.3	0.3	0.0	1.0	1.7	0.0	0.0	3.0	0.3
RB-21	soPl	Fluvial	QW	cu			23.3	10.0	2.0	0.0	0.0	0.3	0.0	0.0	1.3	2.0	0.0	0.0	8.7	0.3
RB-22	soPl	Fluvial	QW	cu			28.7	10.0	0.3	0.0	0.0	0.7	0.3	0.0	1.3	1.7	0.0	0.0	3.7	0.0
RB-23	soPl	Fluvial	QW	cu			30.0	9.0	0.0	0.0	0.0	0.3	1.0	0.0	0.3	3.3	0.0	0.3	1.3	0.0
RB-24	soPl	Fluvial	QW	pb			27.3	9.3	0.0	0.0	0.0	0.3	0.3	0.0	1.7	1.3	0.0	0.0	2.7	0.0
RB-25	soPl	Fluvial	QW	cu			33.7	9.3	0.0	0.0	0.0	0.3	0.3	0.0	0.0	1.7	0.0	0.0	0.7	0.0
RB-26	soPl	Fluvial	QW	pb			39.3	8.7	1.0	0.7	0.3	0.0	1.7	0.3	0.0	0.0	0.0	0.0	0.7	0.3
RB-27	soPl	Fluvial	QW	cu			32.3	10.7	0.3	0.7	0.3	0.3	0.3	1.0	0.0	0.3	0.0	0.0	1.3	0.7
RB-28	soPl	Fluvial	QW	pb			37.3	7.7	1.3	1.0	0.0	0.0	1.0	0.3	0.0	0.3	0.7	0.3	1.0	0.3
RB-29	soPl	Fluvial	QW	cu			34.3	8.0	0.7	0.7	0.0	0.0	2.0	1.0	0.0	0.3	0.3	0.0	1.3	0.7

ub: unbleached, b: bleached, soPl: Plattensandstein Formation, MB: Mined block, QW: Quarry wall, cu: completely unbleached, fb: fully bleached, pb: partially bleached

Table A.1: (continued).

Sample Name	Accessories	FeOx grain	Deformed clayey clast	Clay undifferentiated	*Authigenic components [%]*	Quartz cement	Carbonate pore-filling	Illitic pore-filling	Illitic pore-lining	Kaolinite undifferentiated	Alkali feldspar cement	diag. TiOx	Illite replacing Kspar	Chlorite replacing undifferentiated	Carbonate replacing undifferentiated	*Spectral gamma ray SGR*	K [%]	U [ppm]	Th [ppm]	SGR [API]
RB-01	0.3	12.7	0.0	0.7		10.0	0.0	1.7	12.7	0.0	4.3	0.7	3.0	0.0	0.0		N/A	N/A	N/A	N/A
RB-02	0.0	9.7	0.0	0.7		12.7	0.0	1.3	8.7	0.0	5.3	0.3	3.3	0.0	0.0		N/A	N/A	N/A	N/A
RB-03	0.7	14.3	0.3	0.7		12.3	0.0	1.0	11.3	0.0	5.0	0.3	0.7	0.0	0.0		N/A	N/A	N/A	N/A
RB-04	0.2	8.3	0.5	1.7		13.5	0.0	1.0	7.8	0.0	4.7	0.0	0.2	0.0	0.0		N/A	N/A	N/A	N/A
RB-05	0.3	13.0	0.0	0.7		17.0	0.3	0.7	3.0	0.0	4.7	0.0	1.3	0.3	0.0		N/A	N/A	N/A	N/A
RB-06	0.0	0.3	0.0	1.0		17.7	0.0	0.7	11.3	0.0	9.7	0.0	2.0	0.0	0.0		N/A	N/A	N/A	N/A
RB-07	0.3	0.0	0.7	1.0		16.0	0.0	1.0	11.7	0.0	9.0	0.0	1.0	0.0	0.0		N/A	N/A	N/A	N/A
RB-08	0.0	12.7	0.0	0.3		16.3	0.7	1.0	4.0	0.0	7.7	0.0	1.3	0.0	0.0		4.8	2.8	12.6	150.5
RB-09	0.0	13.7	1.0	0.0		16.7	0.0	1.0	6.3	0.0	5.7	0.0	0.0	0.0	0.0		5.2	2.6	11.0	149.1
RB-10	0.7	16.7	0.7	1.7		18.3	1.3	0.0	4.0	0.0	4.0	0.3	0.7	0.0	0.0		4.8	2.6	10.3	139.8
RB-11	0.0	15.7	0.3	1.0		19.7	0.0	1.0	5.3	0.0	5.0	0.0	0.3	0.0	0.0		4.9	3.0	10.6	145.9
RB-12	0.3	18.0	1.0	2.0		12.3	0.7	0.3	6.7	0.0	4.0	0.0	0.0	0.7	0.0		5.3	2.7	10.8	150.8
RB-13	0.3	13.0	0.7	1.3		18.7	0.0	2.0	7.0	0.0	5.0	0.0	2.0	0.0	0.0		5.2	2.3	11.1	147.1
RB-14	0.3	11.3	0.0	1.3		20.0	0.0	1.3	6.0	0.0	5.3	0.0	1.0	0.0	0.0		4.9	1.9	10.3	135.8
RB-15	1.0	13.7	1.0	0.3		16.3	0.0	1.0	6.3	0.0	4.3	0.3	2.0	0.3	0.0		5.0	2.3	10.8	142.7
RB-16	0.0	16.3	0.0	1.3		14.3	0.7	0.7	5.0	0.0	3.3	0.0	0.3	0.0	0.0		4.7	2.2	11.2	138.5
RB-17	N/A	N/A	N/A	N/A		N/A	N/A	N/A	N/A	N/A	N/A	N/A	N/A	N/A	N/A		5.0	3.5	14.5	166.9
RB-18	0.3	13.0	0.7	0.7		12.3	0.7	0.7	10.0	0.0	7.0	0.0	2.7	0.0	0.0		4.9	2.4	10.8	141.8
RB-19	N/A	N/A	N/A	N/A		N/A	N/A	N/A	N/A	N/A	N/A	N/A	N/A	N/A	N/A		5.1	3.0	13.7	161.3
RB-20	0.0	15.7	0.0	0.7		15.0	0.0	3.0	6.7	0.0	4.7	0.0	1.7	0.3	0.0		5.1	3.3	13.3	162.2
RB-21	0.3	16.7	1.3	0.0		7.3	0.0	1.7	14.7	0.0	6.7	0.3	1.7	0.0	0.0		5.1	3.1	15.5	169.2
RB-22	0.0	15.3	0.0	0.0		15.3	0.3	2.7	6.3	0.0	7.3	0.0	1.7	0.0	0.0		N/A	N/A	N/A	N/A
RB-23	0.0	15.0	0.7	0.7		17.7	0.3	0.3	4.0	0.0	6.0	0.3	1.7	0.0	0.0		N/A	N/A	N/A	N/A
RB-24	0.0	18.7	1.7	1.3		13.3	0.0	1.3	8.3	0.0	6.7	0.3	0.3	0.0	0.0		N/A	N/A	N/A	N/A
RB-25	0.0	13.7	0.3	0.3		20.0	0.0	0.3	4.0	0.0	5.0	0.7	1.3	0.0	0.0		N/A	N/A	N/A	N/A
RB-26	0.7	6.7	0.7	1.7		12.3	0.0	1.3	7.3	0.0	5.3	0.0	0.7	0.0	0.0		N/A	N/A	N/A	N/A
RB-27	0.0	11.7	0.7	2.7		16.7	0.3	0.0	2.3	0.0	4.0	0.0	1.7	0.0	0.0		N/A	N/A	N/A	N/A
RB-28	0.3	6.0	0.7	1.7		12.7	0.7	0.7	9.3	0.3	6.0	0.0	2.0	0.3	0.0		N/A	N/A	N/A	N/A
RB-29	0.0	13.0	0.7	0.3		17.0	0.3	0.0	2.3	0.0	4.7	0.0	0.7	0.0	0.0		N/A	N/A	N/A	N/A

Table A.1: (continued).

Sample Name	*Porosity [%]*	Intergranular porosity	Secondary porosity - Alkali feldspar	Secondary porosity - rock fragment	Secondary porosity - undifferentiated	Fracture porosity	Moldic ductile lithic pore	*Additional sample information*	Plug porosity [vol%]	Plug permeability [mD]	Point count porosity incl. microporosity [vol%]	Mean grain size [mm]	Grain size class (Udden, 1914 & Wentworth, 1922)	Sorting class (Folk & Ward, 1957)
RB-01		5.3	0.0	0.0	0.3	0.0	0.0		10.8	0.020	16.9	0.131	vfg	ms
RB-02		4.0	0.0	0.0	0.0	0.0	0.0		11.0	0.017	12.7	0.146	fg	ms
RB-03		4.0	0.0	0.0	0.0	0.0	0.0		11.2	0.021	12.7	0.126	vfg	ms
RB-04		5.2	0.0	0.0	0.0	0.0	0.0		N/A	N/A	12.0	0.164	fg	ms
RB-05		5.3	0.0	0.0	0.0	0.0	0.0		22.0	502.889	9.2	0.169	fg	ms
RB-06		2.3	0.0	0.0	0.0	0.0	0.0		17.9	40.025	11.7	0.176	fg	ws
RB-07		5.3	0.3	0.0	0.7	0.0	0.0		N/A	N/A	15.8	0.182	fg	ms
RB-08		7.0	0.0	0.0	0.0	0.0	0.0		18.0	37.902	11.2	0.143	fg	ms
RB-09		3.3	0.0	0.0	0.3	0.0	0.0		18.1	25.478	8.8	0.157	fg	ms
RB-10		6.7	0.0	0.0	0.3	0.0	0.0		17.8	19.833	11.2	0.141	vfg	ms
RB-11		4.3	0.0	0.0	0.0	0.0	0.0		17.7	7.561	9.2	0.153	fg	ms
RB-12		5.0	0.0	0.0	0.0	0.0	0.0		15.7	3.704	11.3	0.166	fg	ms
RB-13		5.0	0.3	0.0	0.0	0.0	0.0		15.2	1.357	13.3	0.161	fg	ms
RB-14		4.7	0.0	0.0	0.0	0.0	0.0		15.6	9.608	10.6	0.172	fg	ms
RB-15		2.3	0.0	0.0	0.0	0.0	0.0		15.3	4.400	9.1	0.140	fg	ms
RB-16		6.7	0.0	0.0	0.0	0.0	0.0		16.1	12.594	11.1	0.151	fg	ms
RB-17		N/A	N/A	N/A	N/A	N/A	N/A		14.7	2.128	N/A	N/A	N/A	N/A
RB-18		6.3	0.7	0.0	0.0	0.0	0.0		16.2	6.248	16.1	0.148	fg	ms
RB-19		N/A	N/A	N/A	N/A	N/A	N/A		10.9	0.038	N/A	N/A	N/A	N/A
RB-20		2.7	0.0	0.0	0.0	0.0	0.0		13.4	0.268	10.3	0.153	fg	ms
RB-21		1.3	0.0	0.0	0.0	0.0	0.0		10.5	0.017	13.4	0.113	vfg	ms
RB-22		3.7	0.0	0.0	0.3	0.3	0.0		15.1	1.204	11.1	0.139	fg	ms
RB-23		7.0	0.3	0.0	0.3	0.0	0.0		16.4	12.462	12.1	0.156	fg	ms
RB-24		4.3	0.3	0.0	0.3	0.0	0.0		13.3	0.342	12.9	0.145	fg	ms
RB-25		8.0	0.3	0.0	0.0	0.0	0.0		16.3	15.542	12.3	0.147	fg	ms
RB-26		10.3	0.0	0.0	0.0	0.0	0.0		14.5	3.933	17.4	0.125	vfg	ms
RB-27		8.7	0.3	0.0	0.3	1.3	1.0		17.9	26.351	15.9	0.145	fg	ms
RB-28		7.7	0.3	0.0	0.0	0.0	0.0		14.6	2.149	17.1	0.142	fg	ms
RB-29		10.7	1.0	0.0	0.0	0.0	0.0		18.8	48.431	14.1	0.166	fg	ms

vfg: very fine-grained sandstone, fg: fine-grained sandstone, ms: moderately sorted, ws: well sorted

Table A.1: (continued).

Sample Name	*Compaction state* IGV [vol%]	COPL [vol%]	CEPL [vol%]	*Grain contact types [%]* none	point contact	long contact	concavo-convex contact	sutured contact	contact strength K_0
RB-01	34.7	15.8	24.7	N/A	N/A	N/A	N/A	N/A	N/A
RB-02	32.3	18.7	23.0	N/A	N/A	N/A	N/A	N/A	N/A
RB-03	34.0	16.7	25.0	N/A	N/A	N/A	N/A	N/A	N/A
RB-04	32.2	18.9	21.9	N/A	N/A	N/A	N/A	N/A	N/A
RB-05	31.33	19.9	20.6	N/A	N/A	N/A	N/A	N/A	N/A
RB-06	41.7	5.7	37.1	N/A	N/A	N/A	N/A	N/A	N/A
RB-07	43.0	3.5	36.3	N/A	N/A	N/A	N/A	N/A	N/A
RB-08	36.7	13.2	25.8	N/A	N/A	N/A	N/A	N/A	N/A
RB-09	33.0	17.9	24.4	N/A	N/A	N/A	N/A	N/A	N/A
RB-10	34.7	15.8	23.6	N/A	N/A	N/A	N/A	N/A	N/A
RB-11	35.3	14.9	26.4	N/A	N/A	N/A	N/A	N/A	N/A
RB-12	29.0	22.5	18.6	N/A	N/A	N/A	N/A	N/A	N/A
RB-13	37.7	11.8	28.8	N/A	N/A	N/A	N/A	N/A	N/A
RB-14	37.3	12.2	28.7	N/A	N/A	N/A	N/A	N/A	N/A
RB-15	30.7	20.7	22.5	N/A	N/A	N/A	N/A	N/A	N/A
RB-16	30.7	20.7	19.0	N/A	N/A	N/A	N/A	N/A	N/A
RB-17	N/A	N/A	N/A	N/A	N/A	N/A	N/A	N/A	N/A
RB-18	37.0	12.7	26.8	N/A	N/A	N/A	N/A	N/A	N/A
RB-19	N/A	N/A	N/A	N/A	N/A	N/A	N/A	N/A	N/A
RB-20	32.0	19.1	23.7	N/A	N/A	N/A	N/A	N/A	N/A
RB-21	32.0	19.1	24.8	N/A	N/A	N/A	N/A	N/A	N/A
RB-22	36.0	14.1	27.5	N/A	N/A	N/A	N/A	N/A	N/A
RB-23	35.7	14.5	24.5	N/A	N/A	N/A	N/A	N/A	N/A
RB-24	34.3	16.2	25.1	N/A	N/A	N/A	N/A	N/A	N/A
RB-25	38.0	11.3	26.6	N/A	N/A	N/A	N/A	N/A	N/A
RB-26	36.7	13.2	22.9	N/A	N/A	N/A	N/A	N/A	N/A
RB-27	33.3	17.5	19.3	N/A	N/A	N/A	N/A	N/A	N/A
RB-28	37.3	12.2	26.0	N/A	N/A	N/A	N/A	N/A	N/A
RB-29	35.0	15.4	20.6	N/A	N/A	N/A	N/A	N/A	N/A

Table A.1: (continued).

Sample Name	Lithostratigraphic unit	Facies	Sample type	Bleaching type	*Point counts*	*Detrital components [%]*	Quartz	Alkali feldspar	Shale RF	Siltstone RF	Sandstone RF	Chert undifferentiated	Silicic VRF	altered Volcanic RF	Phyllite	Metamorphic RF undifferentiated	Plutonic RF undifferentiated	Lithoclast undifferentiated	Mica	Sericite
RB-30	soPl	Fluvial	QW	cu			39.7	6.0	1.0	0.3	0.0	0.0	2.7	0.0	0.0	0.0	0.0	0.0	1.7	0.3
RB-31	soPl	Fluvial	QW	cu			39.7	8.3	0.3	0.0	0.0	0.7	1.7	0.7	0.0	2.0	0.0	0.0	0.3	0.0
RB-32	soPl	Fluvial	QW	cu			39.7	8.0	0.0	0.3	0.0	0.7	0.3	0.3	0.0	0.7	0.0	0.0	0.7	0.0
RB-33	soPl	Fluvial	QW	cu			38.0	10.7	0.7	0.0	0.0	0.0	1.0	0.0	0.0	0.7	0.0	0.0	0.3	0.3
RB-34	soPl	Fluvial	QW	cu			40.3	6.3	0.0	0.3	0.0	0.0	0.3	0.3	0.0	1.3	0.0	0.0	4.0	0.3
RB-35	soPl	Fluvial	QW	pb			40.3	6.3	0.0	0.3	0.0	0.3	0.7	0.0	0.0	2.0	0.0	0.0	2.0	0.0
RB-36	soPl	Fluvial	QW	cu			34.0	8.3	0.3	0.0	0.0	1.3	1.0	0.3	0.0	0.3	0.0	0.0	1.0	0.7
RB-37	soPl	Fluvial	QW	cu			33.3	6.0	0.3	0.3	0.0	0.0	0.0	0.0	0.0	0.7	0.3	0.0	3.0	0.3
RB-38	soPl	Fluvial	QW	pb			34.0	8.7	0.3	0.3	0.0	0.3	0.3	0.0	0.0	1.3	0.0	0.0	3.3	0.0
RB-39	soPl	Fluvial	QW	cu			32.3	7.0	0.3	0.0	0.0	0.7	0.3	0.0	1.0	1.7	0.3	0.0	1.7	0.0
RB-40	soPl	Fluvial	QW	cu			32.3	6.3	0.0	0.3	0.0	0.3	0.0	0.0	0.3	0.0	0.0	0.0	6.0	0.0
RB-41	soPl	Fluvial	QW	cu			31.7	8.3	1.0	0.0	0.0	0.0	0.7	0.0	0.3	1.0	0.0	0.0	3.3	0.0
RB-42	soPl	Fluvial	QW	pb			37.3	8.7	0.3	0.0	0.0	0.7	0.7	0.0	0.3	1.3	0.0	0.0	4.3	0.0
RB-43	soPl	Fluvial	QW	cu			39.3	8.7	2.3	0.0	0.0	0.7	1.0	0.0	0.3	0.7	0.0	0.0	1.0	0.0
RB-04ub	soPl	Fluvial	QW				39.0	11.0	0.7	0.7	0.7	0.0	0.0	0.0	1.0	1.7	0.0	0.0	1.3	0.3
RB-04b	soPl	Fluvial	QW				36.0	12.3	2.0	1.3	1.0	0.7	2.0	0.0	1.3	0.3	0.0	0.0	0.3	0.3
RB-05ub	soPl	Fluvial	QW				43.3	8.7	0.3	0.0	0.7	0.7	0.0	0.0	0.0	1.7	0.0	0.0	1.0	0.3
RB-05b	soPl	Fluvial	QW				36.0	12.3	0.3	0.3	0.0	0.0	2.3	0.0	0.7	1.7	0.3	0.0	0.0	0.0
RB-06	soPl	Fluvial	MB	fb			38.0	9.3	0.3	0.3	0.0	0.3	1.3	0.0	2.0	1.7	0.0	0.0	1.7	0.0
RB-07	soPl	Fluvial	MB	fb			38.3	9.0	1.0	0.3	0.0	0.3	1.0	0.0	1.7	0.3	0.0	0.0	1.0	0.0
RB-17ub	soPl	Fluvial	QW				43.3	10.7	0.7	0.3	0.0	0.3	0.0	0.0	2.0	2.0	0.7	0.0	1.7	0.0
RB-17b	soPl	Fluvial	QW				46.0	10.0	0.0	0.0	0.0	0.0	0.3	0.0	1.3	6.0	0.0	0.0	1.3	0.0
RB-24ub	soPl	Fluvial	QW				27.0	7.3	1.0	0.0	0.0	0.7	0.0	0.0	1.3	0.7	0.0	0.0	8.0	1.7
RB-24b	soPl	Fluvial	QW				36.7	16.0	0.0	0.0	0.0	0.0	2.0	0.0	0.7	4.0	0.0	0.0	1.0	0.0
RB-26ub	soPl	Fluvial	QW				27.0	8.3	1.3	0.0	0.0	1.0	4.3	0.0	0.0	0.3	0.0	0.0	3.3	0.3
RB-26b	soPl	Fluvial	QW				41.3	7.7	0.7	0.0	0.0	0.0	1.3	0.0	0.7	0.7	0.0	0.0	2.0	0.7
RB-28ub	soPl	Fluvial	QW				31.3	5.7	2.3	0.0	0.7	1.0	0.3	0.0	0.3	1.7	0.0	0.0	3.7	0.7
RB-28b	soPl	Fluvial	QW				28.3	14.3	0.7	0.0	0.7	1.0	0.3	0.0	0.3	2.0	0.0	0.0	1.3	0.3

ub: unbleached, b: bleached, soPl: Plattensandstein Formation, MB: Mined block, QW: Quarry wall, cu: completely unbleached, fb: fully bleached, pb: partially bleached

Table A.1: (continued).

Sample Name	Accessories	FeOx grain	Deformed clayey clast	Clay undifferentiated	*Authigenic components [%]*	Quartz cement	Carbonate pore-filling	Illitic pore-filling	Illitic pore-lining	Kaolinite undifferentiated	Alkali feldspar cement	diag. TiOx	Illite replacing Kspar	Chlorite replacing undifferentiated	Carbonate replacing undifferentiated	*Spectral gamma ray SGR*	K [%]	U [ppm]	Th [ppm]	SGR [API]
RB-30	0.0	18.0	0.7	1.3		12.7	0.3	0.7	2.3	0.0	2.3	0.0	1.0	0.0	0.0		N/A	N/A	N/A	N/A
RB-31	0.3	14.3	0.7	1.3		13.3	0.0	0.7	2.7	0.0	3.7	0.0	1.3	0.0	0.0		N/A	N/A	N/A	N/A
RB-32	0.0	11.7	0.7	1.0		16.0	0.0	1.0	3.7	0.0	3.0	0.0	0.7	0.0	0.0		N/A	N/A	N/A	N/A
RB-33	0.3	10.0	0.0	0.7		19.0	0.0	0.7	3.7	0.0	3.7	0.0	1.3	0.3	0.0		N/A	N/A	N/A	N/A
RB-34	0.3	11.3	0.7	2.0		13.3	0.0	1.0	8.3	0.0	3.3	0.0	1.3	0.7	0.0		N/A	N/A	N/A	N/A
RB-35	0.0	11.0	0.0	2.0		13.3	0.3	1.3	6.3	0.0	3.7	0.0	1.0	0.3	0.0		N/A	N/A	N/A	N/A
RB-36	0.0	15.7	0.3	1.7		17.0	0.0	1.3	5.3	0.0	3.7	0.0	1.7	0.0	0.0		N/A	N/A	N/A	N/A
RB-37	0.3	19.3	0.7	2.3		20.0	0.0	1.0	5.7	0.0	3.0	0.0	0.0	0.0	0.0		N/A	N/A	N/A	N/A
RB-38	0.3	19.0	1.0	3.7		9.0	0.0	3.3	7.3	0.3	3.0	0.0	1.0	0.7	0.0		N/A	N/A	N/A	N/A
RB-39	0.0	21.0	0.3	4.3		13.3	0.0	0.3	4.0	0.0	4.0	0.3	0.0	1.0	0.0		N/A	N/A	N/A	N/A
RB-40	0.0	16.7	0.7	2.7		15.7	0.0	1.0	2.7	0.3	6.3	0.3	1.0	0.3	0.0		N/A	N/A	N/A	N/A
RB-41	0.0	17.0	0.3	2.7		14.0	0.0	0.7	7.0	0.0	5.0	0.0	0.0	0.0	0.0		N/A	N/A	N/A	N/A
RB-42	0.3	13.0	1.3	3.0		10.7	0.0	0.0	4.7	0.0	3.3	0.3	1.0	0.3	0.0		N/A	N/A	N/A	N/A
RB-43	0.7	14.3	0.3	2.0		11.0	0.3	0.7	2.7	0.0	5.0	0.0	1.3	0.0	0.0		N/A	N/A	N/A	N/A
RB-04ub	0.0	16.7	0.3	1.7		11.3	0.0	0.3	5.7	0.0	4.0	0.0	0.3	0.0	0.0		N/A	N/A	N/A	N/A
RB-04b	0.3	0.0	0.7	1.7		15.7	0.0	1.7	10.0	0.0	5.3	0.0	0.0	0.0	0.0		N/A	N/A	N/A	N/A
RB-05ub	0.0	13.3	0.0	0.3		11.0	0.0	0.7	2.7	0.0	5.7	0.0	1.7	0.0	0.0		N/A	N/A	N/A	N/A
RB-05b	0.0	0.0	0.0	2.3		21.7	0.0	0.3	8.7	0.0	8.0	0.0	0.3	0.0	0.0		N/A	N/A	N/A	N/A
RB-06	0.0	0.3	0.0	1.0		17.7	0.0	0.7	11.3	0.0	9.7	0.0	2.0	0.0	0.0		N/A	N/A	N/A	N/A
RB-07	0.3	0.0	0.7	1.0		16.0	0.0	1.0	11.7	0.0	9.0	0.0	1.0	0.0	0.0		N/A	N/A	N/A	N/A
RB-17ub	0.7	18.7	0.0	1.3		4.7	0.0	0.3	5.7	0.0	1.7	0.0	3.7	0.0	0.0		N/A	N/A	N/A	N/A
RB-17b	0.3	0.0	0.3	0.3		12.0	0.0	0.0	6.0	0.0	6.0	0.0	2.7	0.0	0.0		N/A	N/A	N/A	N/A
RB-24ub	0.3	25.0	3.0	1.0		11.7	0.3	0.3	4.0	0.0	4.3	0.0	0.3	0.0	0.0		N/A	N/A	N/A	N/A
RB-24b	0.0	0.0	0.3	0.7		16.0	0.0	0.3	3.3	0.0	9.0	0.0	5.0	0.0	0.0		N/A	N/A	N/A	N/A
RB-26ub	0.0	15.3	3.0	1.0		17.7	0.0	0.7	10.0	0.0	2.7	0.3	1.0	0.0	0.0		N/A	N/A	N/A	N/A
RB-26b	0.3	4.0	0.3	1.7		16.0	1.3	0.3	4.7	0.0	4.7	0.3	0.7	0.0	0.0		N/A	N/A	N/A	N/A
RB-28ub	0.0	13.7	1.3	1.7		15.3	0.0	0.3	8.7	0.0	5.0	0.0	0.7	1.3	0.0		N/A	N/A	N/A	N/A
RB-28b	0.7	0.7	0.0	1.3		16.0	2.7	2.0	6.7	0.0	7.3	0.7	2.3	0.3	0.3		N/A	N/A	N/A	N/A

Table A.1: (continued).

Sample Name	*Porosity [%]*						*Additional sample information*					
	Intergranular porosity	Secondary porosity - Alkali feldspar	Secondary porosity - rock fragment	Secondary porosity - undifferentiated	Fracture porosity	Moldic ductile lithic pore	Plug porosity [vol%]	Plug permeability [mD]	Point count porosity incl. microporosity [vol%]	Mean grain size [mm]	Grain size class (Udden, 1914 & Wentworth, 1922)	Sorting class (Folk & Ward, 1957)
RB-30	6.7	1.3	0.0	0.7	0.0	0.3	15.6	8.350	12.6	0.159	fg	ms
RB-31	6.3	0.7	0.0	1.0	0.0	0.0	16.1	7.368	12.0	0.139	vfg	ms
RB-32	11.0	0.3	0.0	0.3	0.0	0.0	16.6	11.355	15.9	0.162	fg	ms
RB-33	7.0	1.0	0.0	0.7	0.0	0.0	18.5	31.250	12.8	0.169	fg	ms
RB-34	3.3	0.0	0.0	1.0	0.0	0.0	14.5	0.223	12.8	0.120	vfg	ms
RB-35	8.3	0.0	0.0	0.3	0.0	0.0	15.8	5.290	15.3	0.132	vfg	ms
RB-36	5.3	0.0	0.0	0.7	0.0	0.0	13.1	0.284	12.3	0.160	fg	ms
RB-37	3.3	0.0	0.0	0.0	0.0	0.0	16.0	3.091	9.1	0.138	fg	ms
RB-38	2.7	0.0	0.0	0.0	0.0	0.0	13.8	0.196	12.9	0.137	vfg	ms
RB-39	4.7	0.3	0.0	1.0	0.0	0.0	15.0	0.851	11.7	0.150	fg	ws
RB-40	6.3	0.0	0.0	0.3	0.0	0.0	15.1	0.570	11.7	0.141	fg	ms
RB-41	5.3	0.3	0.0	1.0	0.0	0.3	15.2	0.767	13.4	0.147	fg	ms
RB-42	5.0	2.3	0.0	1.0	0.0	0.0	15.1	0.475	14.3	0.136	vfg	ms
RB-43	6.7	0.7	0.0	0.3	0.0	0.0	16.9	19.287	11.8	0.159	fg	ms
RB-04ub	3.3	0.0	0.0	0.0	0.0	0.0	N/A	N/A	8.4	0.163	fg	ms
RB-04b	7.0	0.0	0.0	0.0	0.0	0.0	N/A	N/A	15.6	0.179	fg	ms
RB-05ub	8.0	0.0	0.0	0.0	0.0	0.0	N/A	N/A	11.3	0.157	fg	ms
RB-05b	4.7	0.0	0.0	0.0	0.0	0.0	N/A	N/A	11.8	0.185	fg	ws
RB-06	2.3	0.0	0.0	0.0	0.0	0.0	N/A	N/A	11.7	0.176	fg	ws
RB-07	5.3	0.3	0.0	0.7	0.0	0.0	N/A	N/A	15.8	0.182	fg	ms
RB-17ub	1.7	0.0	0.0	0.0	0.0	0.0	N/A	N/A	8.5	0.132	vfg	ms
RB-17b	3.3	0.3	0.0	3.7	0.0	0.0	N/A	N/A	13.1	0.165	fg	ms
RB-24ub	2.0	0.0	0.0	0.0	0.0	0.0	N/A	N/A	7.0	0.135	vfg	ms
RB-24b	3.0	1.3	0.0	0.7	0.0	0.0	N/A	N/A	11.0	0.175	fg	ms
RB-26ub	2.3	0.0	0.0	0.0	0.0	0.0	N/A	N/A	11.8	0.157	fg	ms
RB-26b	9.0	0.7	0.0	1.0	0.0	0.0	N/A	N/A	15.3	0.182	fg	ms
RB-28ub	3.0	0.3	0.0	1.0	0.0	0.0	N/A	N/A	12.7	0.146	fg	ms
RB-28b	8.0	0.7	0.3	0.7	0.0	0.0	N/A	N/A	17.5	0.162	fg	ms

vfg: very fine-grained sandstone, fg: fine-grained sandstone, ms: moderately sorted, ws: well sorted

Table A.1: (continued).

Sample Name	*Compaction state*	IGV [vol%]	COPL [vol%]	CEPL [vol%]	*Grain contact assessment [vol%]*	none	point contact	long contact	concavo-convex contact	sutured contact	contact strength K_0
RB-30		25.0	26.7	13.4		N/A	N/A	N/A	N/A	N/A	N/A
RB-31		26.7	25.0	15.3		N/A	N/A	N/A	N/A	N/A	N/A
RB-32		34.7	15.8	19.9		N/A	N/A	N/A	N/A	N/A	N/A
RB-33		34.0	16.7	22.5		N/A	N/A	N/A	N/A	N/A	N/A
RB-34		29.3	22.2	20.2		N/A	N/A	N/A	N/A	N/A	N/A
RB-35		33.3	17.5	20.6		N/A	N/A	N/A	N/A	N/A	N/A
RB-36		32.7	18.3	22.3		N/A	N/A	N/A	N/A	N/A	N/A
RB-37		33.0	17.9	24.4		N/A	N/A	N/A	N/A	N/A	N/A
RB-38		25.7	26.0	17.0		N/A	N/A	N/A	N/A	N/A	N/A
RB-39		26.7	25.0	16.5		N/A	N/A	N/A	N/A	N/A	N/A
RB-40		32.7	18.3	21.5		N/A	N/A	N/A	N/A	N/A	N/A
RB-41		32.0	19.1	21.6		N/A	N/A	N/A	N/A	N/A	N/A
RB-42		24.0	27.6	13.8		N/A	N/A	N/A	N/A	N/A	N/A
RB-43		26.3	25.3	14.7		N/A	N/A	N/A	N/A	N/A	N/A
RB-04ub		24.7	27.0	15.6		0	2	21	70	7	2.8
RB-04b		39.7	8.8	29.8		2	21	52	25	0	2.0
RB-05ub		28.0	23.6	15.3		2	2	41	54	1	2.6
RB-05b		43.3	2.9	37.5		7	32	42	18	1	1.9
RB-06		41.7	5.7	37.1		2	23	45	30	0	2.1
RB-07		43.0	3.5	36.3		5	23	45	26	1	2.1
RB-17ub		14.0	36.0	7.9		0	0	43	54	3	2.6
RB-17b		27.3	24.3	18.2		4	15	48	33	0	2.2
RB-24ub		22.7	28.9	14.7		0	0	36	61	3	2.7
RB-24b		31.67	19.5	23.1		4	28	49	19	0	1.9
RB-26ub		33.7	17.1	26.0		0	2	38	56	4	2.6
RB-26b		36.3	13.6	23.6		7	29	41	23	0	1.9
RB-28ub		32.3	18.7	23.8		0	2	47	46	5	2.5
RB-28b		43.3	2.9	34.3		12	16	47	25	0	2.1

Table A.1: (continued).

Sample Name	Lithostratigraphic unit	Facies	Sample type	Bleaching type	*Point counts*	*Detrital components [%]*	Quartz	Alkali feldspar	Shale RF	Siltstone RF	Sandstone RF	Chert undifferentiated	Silicic VRF	altered Volcanic RF	Phyllite	Metamorphic RF undifferentiated	Plutonic RF undifferentiated	Lithoclast undifferentiated	Mica	Sericite
RB-35ub	soPl	Fluvial	QW				34.0	4.0	0.7	0.0	0.3	0.7	0.0	0.0	0.3	2.0	0.0	0.0	4.0	0.0
RB-35b	soPl	Fluvial	QW				41.0	8.3	1.3	0.0	0.0	0.0	0.7	0.0	0.7	0.0	0.0	0.0	2.7	0.3
RB-38ub	soPl	Fluvial	QW				36.3	7.0	0.3	0.0	0.0	0.0	0.3	0.0	1.0	0.3	0.0	0.0	3.3	0.7
RB-38b	soPl	Fluvial	QW				37.7	7.3	2.3	0.3	0.0	0.3	0.7	0.0	0.3	0.0	0.0	0.0	1.0	0.3
RB-42ub	soPl	Fluvial	QW				38.3	6.7	1.3	0.0	0.3	0.0	0.3	0.0	0.3	0.7	0.0	0.0	3.3	0.3
RB-42b	soPl	Fluvial	QW				37.3	11.0	0.7	0.0	0.3	0.3	0.3	0.0	0.0	0.0	0.0	0.0	3.0	0.3

ub: unbleached, b: bleached, soPl: Plattensandstein Formation, MB: Mined block, QW: Quarry wall, cu: completely unbleached, fb: fully bleached, pb: partially bleached

Table A.1: (continued).

Sample Name	Accessories	FeOx grain	Deformed clayey clast	Clay undifferentiated	*Authigenic components [%]*	Quartz cement	Carbonate pore-filling	Illitic pore-filling	Illitic pore-lining	Kaolinite undifferentiated	Alkali feldspar cement	diag. TiOx	Illite replacing Kspar	Chlorite replacing undifferentiated	Carbonate replacing undifferentiated	*Spectral gamma ray SGR*	K [%]	U [ppm]	Th [ppm]	SGR [API]
RB-35ub	0.0	18.0	1.3	1.0		15.7	0.3	0.7	4.7	0.0	1.7	0.0	1.0	0.3	0.0		N/A	N/A	N/A	N/A
RB-35b	0.3	0.0	0.0	2.3		18.3	0.7	2.3	10.3	0.0	2.0	0.3	1.3	0.3	0.0		N/A	N/A	N/A	N/A
RB-38ub	0.3	20.3	2.3	1.3		11.3	0.0	0.0	7.0	0.0	3.0	0.0	0.7	0.0	0.0		N/A	N/A	N/A	N/A
RB-38b	0.3	0.0	0.7	3.0		20.7	0.7	2.3	8.3	0.0	4.7	0.0	2.3	0.0	0.0		N/A	N/A	N/A	N/A
RB-42ub	0.0	14.7	1.0	2.0		13.3	0.0	2.0	6.0	0.0	3.7	0.3	2.0	0.0	0.0		N/A	N/A	N/A	N/A
RB-42b	0.3	0.0	0.7	1.7		22.3	0.0	1.0	7.7	0.0	7.0	0.0	1.0	0.0	0.0		N/A	N/A	N/A	N/A

Table A.1: (continued).

Sample Name	*Porosity [%]* Intergranular porosity	Secondary porosity - Alkali feldspar	Secondary porosity - rock fragment	Secondary porosity - undifferentiated	Fracture porosity	Moldic ductile lithic pore	*Additional sample information* Plug porosity [vol%]	Plug permeability [mD]	Point count porosity incl. microporosity [vol%]	Mean grain size [mm]	Grain size class (Udden, 1914 & Wentworth, 1922)	Sorting class (Folk & Ward, 1957)
RB-35ub	8.3	1.0	0.0	0.0	0.0	0.0	N/A	N/A	14.7	0.152	fg	ms
RB-35b	6.0	0.3	0.3	0.0	0.0	0.0	N/A	N/A	16.9	0.176	fg	ws
RB-38ub	4.0	0.0	0.0	0.7	0.0	0.0	N/A	N/A	11.4	0.130	vfg	ms
RB-38b	6.3	0.0	0.0	0.3	0.0	0.0	N/A	N/A	16.8	0.151	fg	ms
RB-42ub	3.0	0.3	0.0	0.0	0.0	0.0	N/A	N/A	11.2	0.140	fg	ms
RB-42b	4.3	0.7	0.0	0.0	0.0	0.0	N/A	N/A	12.3	0.164	fg	ws

vfg: very fine-grained sandstone, fg: fine-grained sandstone, ms: moderately sorted, ws: well sorted

Table A.1: (continued).

Sample Name	*Compaction state*	IGV [vol%]	COPL [vol%]	CEPL [vol%]	*Grain contact assessment [vol%]*	none	point contact	long contact	concavo-convex contact	sutured contact	contact strength K_0
RB-35ub		31.3	19.9	18.4		0	2	37	58	3	2.6
RB-35b		40.0	8.3	31.2		4	25	51	18	2	2.0
RB-38ub		25.3	26.3	15.7		0	2	37	58	3	2.6
RB-38b		43.0	3.5	35.4		4	25	56	15	0	1.9
RB-42ub		28.3	23.3	19.4		0	3	38	55	4	2.6
RB-42b		42.3	4.6	36.2		5	29	44	22	0	1.9

Table A.2: Compilation of point count, petrophysical data and additional sample information (Chapter 4).

Sample Name	Lithostratigraphic unit	Facies	Lithofacies type after Miall (1996)	Orientation plug's longitudinal axis	Orientation/ Strike sampled quarry wall	*Point counts*	*Detrital components [%]*	Quartz	Alkali feldspar	Shale RF	Siltstone RF	Sandstone RF	Chert undifferentiated	Silicic VRF	Feldspathic VRF	Phyllite	Metamorphic RF undifferentiated	Plutonic RF undifferentiated	Lithoclast undifferentiated	Mica
LK-05	sV	Fluvial	St	hor.	N-S			58.0	5.3	0.0	0.0	0.0	0.0	0.3	0.0	0.0	1.3	0.3	0.0	1.0
LK-06	sV	Fluvial	St	hor.	N-S			60.0	6.3	0.0	0.0	0.0	0.3	0.3	0.0	0.0	1.7	0.0	0.0	0.7
LK-07	sV	Fluvial	St	hor.	N-S			55.0	12.7	0.0	0.0	0.3	0.7	1.0	0.0	0.3	0.7	0.3	0.0	0.3
LK-08	sV	Fluvial	St	hor.	N-S			56.0	7.3	0.0	0.0	0.0	0.7	0.3	0.0	0.0	2.3	0.3	0.0	0.3
LK-09	sV	Fluvial	St	hor.	N-S			50.0	7.0	0.0	0.3	0.0	0.7	0.7	0.0	0.0	1.3	0.3	0.3	0.0
LK-10	sV	Fluvial	St	hor.	N-S			52.3	8.3	0.0	0.0	0.0	0.7	3.7	0.3	0.0	2.3	0.0	0.3	0.0
LK-11	sV	Fluvial	St	hor.	W-E			53.7	9.0	0.0	0.3	0.0	0.3	1.3	0.0	0.0	0.7	0.3	0.0	0.3
LK-12	sV	Fluvial	St	hor.	W-E			52.0	9.0	0.0	0.0	0.7	0.3	1.3	0.3	0.0	0.7	0.0	0.0	0.0
LK-13	sV	Fluvial	St	hor.	W-E			58.3	7.7	0.0	0.0	0.7	0.3	1.0	0.0	0.0	0.0	0.3	0.0	0.0
LK-14	sV	Fluvial	St	hor.	W-E			58.7	8.7	0.0	0.0	0.3	1.0	1.0	0.3	0.0	2.0	0.0	0.3	0.0
LK-15	sV	Fluvial	St	hor.	W-E			62.7	7.3	0.0	0.0	0.3	0.7	2.3	0.0	0.0	1.3	0.0	0.0	0.7
LK-16	sV	Fluvial	St	hor.	W-E			47.0	9.3	0.0	0.0	0.0	2.0	0.7	0.0	0.3	1.7	0.0	0.0	0.3
LK-17	sV	Fluvial	St	hor.	W-E			55.0	5.0	0.0	0.0	0.0	0.0	0.7	0.3	0.0	0.3	0.0	0.0	0.3
LK-18	sV	Fluvial	St	hor.	W-E			54.0	9.0	0.0	0.0	0.0	0.3	1.3	0.0	0.0	1.0	0.0	0.0	0.0
LK-19	sV	Fluvial	St	hor.	W-E			53.7	7.0	0.0	0.0	0.0	0.7	1.0	0.0	0.0	1.3	0.0	0.0	0.0
LK-20	sV	Fluvial	Swr	hor.	W-E			62.7	11.3	0.3	0.0	0.0	0.7	1.3	0.0	0.0	2.0	0.0	0.3	0.0
LK-21	sV	Fluvial	Swr	hor.	W-E			54.7	10.0	0.0	0.0	0.0	0.7	2.3	0.3	0.3	2.3	0.0	0.0	2.0
LK-22	sV	Fluvial	Swr	hor.	W-E			56.7	9.0	0.0	0.0	0.3	0.3	1.7	0.0	0.0	2.0	0.3	0.0	1.0
LK-23	sV	Fluvial	Swr	hor.	W-E			48.7	11.3	0.3	0.0	0.0	0.3	2.0	0.0	0.0	2.0	0.7	0.0	1.3
LK-24	sV	Fluvial	Swr	hor.	W-E			55.7	8.0	0.0	0.0	0.0	0.0	1.3	0.0	0.0	1.0	0.0	0.0	1.0
LK-25	sV	Fluvial	Swr	hor.	W-E			64.0	9.0	0.0	0.3	0.0	0.0	1.0	0.3	0.0	2.3	0.7	0.0	0.0
LK-26	sV	Fluvial	Swr	hor.	N-S			51.7	8.3	0.0	0.0	0.0	1.0	1.7	0.3	0.0	1.3	0.0	0.0	1.0
LK-27	sV	Fluvial	Swr	hor.	N-S			55.3	9.0	0.0	0.0	0.0	0.3	1.3	0.3	0.0	2.3	0.0	0.0	0.3
LK-28	sV	Fluvial	Swr	hor.	N-S			59.7	8.0	0.0	0.0	0.0	0.0	2.7	0.0	0.0	2.3	0.0	0.0	0.7
LK-29	sV	Fluvial	Swr	hor.	N-S			54.7	8.0	0.0	0.0	0.0	0.0	1.7	0.0	0.0	1.3	0.0	0.0	0.0
LK-30	sV	Fluvial	Swr	hor.	N-S			64.0	7.3	0.0	0.0	0.0	0.3	2.3	0.0	0.0	1.0	0.0	0.0	0.3
LK-31	sV	Fluvial	Swr	hor.	N-S			57.3	9.3	0.0	0.0	0.0	0.3	1.3	0.0	0.0	0.7	0.0	0.0	2.7
LK-32	sV	Fluvial	St	vert.	N-S			63.3	11.0	0.0	0.0	0.0	0.0	2.7	0.3	0.0	0.3	0.0	0.0	0.0
LK-33	sV	Fluvial	St	vert.	N-S			59.0	9.3	0.0	0.0	0.0	0.7	1.3	0.0	0.0	0.7	0.3	0.0	0.0
LK-34	sV	Fluvial	St	vert.	N-S			56.3	5.7	0.0	0.0	0.0	0.0	2.0	0.7	0.0	0.3	0.3	0.0	0.0
LK-35	sV	Fluvial	St	vert.	W-E			60.3	5.7	0.0	0.0	0.3	0.7	2.0	1.0	0.0	0.7	0.3	0.0	0.0

sV: Vogesensandstein Formation, St: trough cross-bedded sandstone, Swr: wavy to ripple-bedded sandstone, hor.: horizontal, vert.: vertical

Table A.2: (continued).

				Authigenic components [%]										Porosity [%]				
Sample Name	Accessories	Iron Oxide	Detrital clay	Quartz cement	Illite tangential	Illite radial	Illite meshwork	Alkali feldspar cement	Kaolinite undifferentiated	Carbonate pore-filling	Barite cement	Illite replacing alkali feldspar	Carbonate replacing undifferentiated grain	Intergranular porosity	Secondary porosity - Alkali feldspar	Secondary porosity - Rock fragment	Secondary porosity - undifferentiated	Secondary porosity - after nodule dissolution
LK-05	0.3	10.0	0.3	12.3	1.0	1.0	0.7	0.3	0.0	0.0	0.0	1.0	0.0	5.3	1.0	0.3	0.0	0.0
LK-06	0.0	7.7	0.0	9.7	1.3	3.3	1.0	0.0	0.7	0.0	0.0	2.7	0.0	3.7	0.7	0.0	0.0	0.0
LK-07	0.3	8.3	0.7	6.0	3.0	3.7	1.3	0.7	0.3	0.0	0.0	2.7	0.0	0.7	1.0	0.0	0.0	0.0
LK-08	0.0	4.7	0.0	13.7	1.3	3.7	0.0	0.3	0.3	0.0	0.0	1.3	0.0	5.7	1.3	0.3	0.0	0.0
LK-09	0.0	6.7	0.0	12.3	1.0	6.0	1.3	0.7	0.0	0.0	0.0	3.0	0.0	5.7	2.3	0.0	0.3	0.0
LK-10	0.0	10.0	0.0	9.0	1.0	3.7	0.3	0.3	0.0	0.0	0.0	1.7	0.0	5.0	1.0	0.0	0.0	0.0
LK-11	0.0	4.7	0.3	14.3	2.0	0.0	0.3	0.0	0.7	0.0	0.3	0.0	0.0	9.7	1.7	0.0	0.0	0.0
LK-12	0.0	7.3	0.3	14.0	0.0	0.3	1.3	0.0	0.3	0.0	0.0	1.0	0.0	8.7	2.3	0.0	0.0	0.0
LK-13	0.0	9.0	0.0	12.7	0.3	0.7	0.0	0.7	0.3	0.0	0.0	0.3	0.0	6.7	1.0	0.0	0.0	0.0
LK-14	0.0	6.3	0.3	9.3	1.0	0.7	0.7	0.0	0.0	0.0	0.0	0.0	0.0	6.7	2.3	0.0	0.3	0.0
LK-15	0.0	11.0	0.0	4.0	4.3	1.3	0.7	0.7	0.0	0.0	0.0	0.7	0.0	1.7	0.0	0.0	0.3	0.0
LK-16	0.0	11.3	0.0	10.7	2.0	5.0	1.3	0.0	0.0	0.0	0.0	2.0	0.0	5.7	0.3	0.3	0.0	0.0
LK-17	0.0	11.7	0.0	8.3	3.0	1.7	0.7	0.7	0.0	0.0	0.0	2.0	0.0	7.3	1.7	0.0	0.0	1.3
LK-18	0.3	7.3	0.0	14.0	1.7	0.7	0.0	0.3	0.0	0.0	0.0	1.3	0.0	6.7	1.3	0.7	0.0	0.0
LK-19	0.0	9.0	0.0	11.3	2.0	0.3	1.0	1.0	0.0	0.0	0.0	1.0	0.0	6.0	1.0	0.0	0.0	3.7
LK-20	0.0	6.0	0.0	0.7	3.3	4.7	1.3	0.0	1.0	0.0	0.0	3.0	0.0	0.3	0.7	0.0	0.0	0.3
LK-21	0.3	17.3	0.0	1.0	3.3	1.3	1.0	0.0	0.0	0.0	0.0	1.0	0.0	0.7	1.0	0.0	0.3	0.0
LK-22	0.7	16.0	0.0	2.3	3.3	1.0	1.0	0.3	0.0	0.0	0.0	1.7	0.0	1.3	0.7	0.0	0.0	0.3
LK-23	0.3	22.7	0.0	1.3	2.3	1.3	0.0	0.0	0.3	0.0	0.0	2.0	0.0	1.3	1.0	0.0	0.7	0.0
LK-24	0.3	15.7	0.0	2.0	2.3	3.3	1.3	0.3	0.0	0.0	0.0	2.3	0.0	3.3	1.7	0.3	0.0	0.0
LK-25	0.0	6.7	0.0	4.3	2.7	3.7	0.3	0.3	0.3	0.0	0.0	1.3	0.0	1.7	1.0	0.0	0.0	0.0
LK-26	0.3	21.0	0.0	0.7	3.7	1.0	0.3	0.0	0.0	0.0	0.0	5.3	0.0	1.3	1.0	0.0	0.0	0.0
LK-27	0.0	21.7	0.0	1.0	2.3	2.3	0.0	0.0	0.3	0.0	0.0	1.3	0.0	0.7	0.3	0.0	0.0	1.0
LK-28	0.3	15.0	0.0	0.0	3.3	1.7	1.0	0.3	0.7	0.0	0.0	2.0	0.0	0.7	0.3	0.0	0.0	1.3
LK-29	0.3	19.3	0.0	1.3	2.0	0.7	0.3	0.3	0.0	0.0	0.0	3.7	0.0	2.0	0.7	0.0	0.7	3.0
LK-30	0.0	10.3	0.0	4.0	1.3	3.7	0.7	0.0	0.0	0.0	0.0	1.0	0.0	2.0	1.0	0.0	0.7	0.0
LK-31	0.3	22.3	0.0	1.0	3.0	0.0	0.0	0.3	0.0	0.0	0.0	1.0	0.0	0.3	0.0	0.0	0.0	0.0
LK-32	0.0	3.7	0.0	7.0	1.3	3.3	0.7	0.0	0.3	0.0	0.0	1.0	0.0	5.0	0.0	0.0	0.0	0.0
LK-33	0.7	11.7	0.0	3.7	3.0	4.7	0.7	0.3	0.7	0.0	0.0	1.7	0.0	0.7	1.0	0.0	0.0	0.0
LK-34	0.0	6.7	0.0	10.3	0.0	6.7	0.0	0.7	0.0	0.0	0.0	1.0	0.0	8.3	1.0	0.0	0.0	0.0
LK-35	0.0	6.7	0.0	12.0	0.0	0.3	0.0	0.3	0.3	0.0	0.0	0.3	0.0	8.7	0.3	0.0	0.0	0.0

Table A.2: (continued).

Sample Name	*Spectral gamma ray SGR*	K [%]	U [ppm]	Th [ppm]	SGR [API]	*Additional sample information*	Plug porosity [vol%]	Plug permeability [mD]	Point count porosity incl. microporosity [vol%]	Mean grain size [mm]	Mean grain size [Phi] (Udden, 1914 & Wentworth, 1922)	Grain size class (Udden, 1914 & Wentworth, 1922)	Sorting class (Folk & Ward, 1957)
LK-05		2.4	0.9	2.9	57.8		16.9	435.862	9.2	0.307	1.881	mg	ms
LK-06		2.7	0.8	4.0	66.3		14.5	21.635	9.9	0.240	2.218	fg	ms
LK-07		2.7	1.2	4.3	70.7		13.0	2.035	8.9	0.179	2.735	fg	ms
LK-08		2.7	1.2	3.5	67.5		17.2	118.108	11.5	0.285	1.994	mg	ms
LK-09		3.1	1.5	5.3	83.6		15.8	33.615	15.5	0.255	2.156	fg	ms
LK-10		3.1	0.6	4.7	73.9		16.7	52.205	10.2	0.255	2.113	fg	ms
LK-11		2.6	1.0	3.0	62.3		18.6	548.833	13.3	0.313	1.763	mg	ms
LK-12		2.6	0.6	3.7	61.8		18.2	772.873	13.0	0.318	1.788	mg	ms
LK-13		2.6	1.2	3.0	63.9		19.2	830.451	8.7	0.273	1.996	mg	ms
LK-14		2.3	1.0	2.9	57.0		18.0	423.693	11.0	0.268	2.024	fg	ms
LK-15		2.4	1.1	3.0	59.9		12.2	10.816	6.4	0.318	1.769	mg	ms
LK-16		3.2	1.1	4.2	77.6		16.2	0.823	12.8	0.184	2.597	fg	ms
LK-17		2.7	1.0	3.1	64.3		19.3	72.982	15.0	0.257	2.068	fg	ms
LK-18		2.6	0.8	3.4	62.3		18.7	371.716	11.0	0.264	1.998	mg	ws
LK-19		3.0	1.2	3.6	72.8		20.9	223.754	13.4	0.238	2.193	fg	ms
LK-20		3.1	1.2	5.3	81.1		12.9	0.416	9.5	0.272	2.017	fg	ms
LK-21		3.2	1.6	4.9	84.4		7.9	0.002	6.2	0.198	2.517	fg	ms
LK-22		3.3	0.9	5.3	82.0		8.2	0.005	6.7	0.243	2.187	fg	ms
LK-23		3.2	1.5	5.1	84.4		8.6	0.003	6.9	0.143	2.997	fg	ms
LK-24		3.4	0.9	6.0	86.3		1.5	0.044	11.2	0.190	2.619	fg	ms
LK-25		3.2	1.3	4.4	80.0		14.6	1.039	7.9	0.229	2.390	fg	ms
LK-26		3.7	1.7	5.5	95.8		12.4	0.038	8.8	0.166	2.757	fg	ms
LK-27		3.5	1.2	7.0	94.3		10.3	0.029	5.9	0.183	2.732	fg	ms
LK-28		3.5	1.2	5.9	90.0		10.3	0.007	7.7	0.186	2.602	fg	ms
LK-29		3.5	1.1	5.8	88.8		11.2	0.019	10.5	0.172	2.754	fg	ms
LK-30		3.3	1.3	5.0	84.0		12.1	0.066	7.9	0.198	2.575	fg	ms
LK-31		3.4	1.4	5.7	89.2		7.3	0.004	2.9	0.125	3.300	v. fg	ms
LK-32		N/A	N/A	N/A	N/A		16.4	1.773	9.1	0.322	1.829	mg	ms
LK-33		N/A	N/A	N/A	N/A		14.5	0.096	8.3	0.193	2.557	fg	ms
LK-34		N/A	N/A	N/A	N/A		14.4	0.002	14.2	0.328	1.703	mg	ms
LK-35		N/A	N/A	N/A	N/A		21.1	777.180	9.6	0.297	1.848	mg	ms

v. fg: very fine-grained, fg: fine-grained, mg: medium-grained, ms: moderately sorted, ws: well sorted

Table A.2: (continued).

Sample Name	*Compaction parameters* IGV [vol%]	COPL [vol%]	CEPL [vol%]	*Grain contact types [%]* none	point contact	long contact	concavo-convex contact	sutured contact
LK-05	20.7	30.7	10.6	3	18	32	46	1
LK-06	19.7	31.5	11.0	0	6	33	51	10
LK-07	16.0	34.5	10.0	0	3	15	77	5
LK-08	25.0	26.7	14.2	6	11	43	40	0
LK-09	27.0	24.7	16.1	3	15	43	39	0
LK-10	19.3	31.8	9.8	1	8	30	61	0
LK-11	27.3	24.3	13.4	3	11	36	50	0
LK-12	24.7	27.0	11.7	3	10	31	56	0
LK-13	21.3	30.1	10.3	9	7	33	51	0
LK-14	18.3	32.7	7.9	4	5	26	62	3
LK-15	12.7	37.0	6.9	2	1	13	79	5
LK-16	24.7	27.0	13.9	0	3	25	66	6
LK-17	21.7	29.8	10.1	2	5	33	60	0
LK-18	23.3	28.3	12.0	0	12	39	49	0
LK-19	21.7	29.8	11.0	4	17	33	43	3
LK-20	11.3	38.0	6.8	0	2	9	78	11
LK-21	7.3	40.6	4.0	0	0	3	92	5
LK-22	9.3	39.3	4.9	0	0	6	87	7
LK-23	6.7	41.1	3.1	0	0	4	91	5
LK-24	12.7	37.0	5.9	0	1	7	88	4
LK-25	13.3	36.5	7.4	0	0	12	80	8
LK-26	7.0	40.9	3.4	0	0	4	94	2
LK-27	6.7	41.1	3.5	0	1	8	84	7
LK-28	7.7	40.4	4.2	0	0	2	84	14
LK-29	6.7	41.1	2.8	0	1	7	86	6
LK-30	11.7	37.7	6.0	0	2	11	86	1
LK-31	4.7	42.3	2.5	0	0	1	95	4
LK-32	17.7	33.2	8.5	1	3	26	69	1
LK-33	13.7	36.3	8.3	0	1	20	79	0
LK-34	26.0	25.7	13.1	6	17	38	39	0
LK-35	21.7	29.8	9.1	10	15	36	38	1

Table A.2: (continued).

Sample Name	Lithostratigraphic unit	Facies	Lithofacies type after Miall (1996)	Orientation plug's longitudinal axis	Orientation/ Strike sampled quarry wall	*Point counts*	*Detrital components [%]*	Quartz	Alkali feldspar	Shale RF	Siltstone RF	Sandstone RF	Chert undifferentiated	Silicic VRF	Feldspathic VRF	Phyllite	Metamorphic RF undifferentiated	Plutonic RF undifferentiated	Lithoclast undifferentiated	Mica
LK-36	sV	Fluvial	St	vert.	W-E			56.0	8.7	0.0	0.0	0.0	0.0	1.0	0.0	0.0	0.3	0.3	0.0	0.0
LK-37	sV	Fluvial	St	vert.	W-E			65.7	6.0	0.0	0.0	0.7	0.0	1.3	0.0	0.0	1.3	0.0	0.0	0.3
LK-38	sV	Fluvial	St	vert.	W-E			43.7	10.7	0.0	0.0	0.3	0.7	1.3	0.0	0.0	2.3	1.0	1.3	0.0
LK-39	sV	Fluvial	St	vert.	W-E			53.0	7.3	0.0	0.0	0.0	1.0	1.7	0.0	0.0	0.7	0.0	0.0	0.0
LK-40	sV	Fluvial	Swr	vert.	W-E			62.0	10.0	0.0	0.0	0.0	1.0	1.3	0.3	0.0	1.0	0.7	0.3	0.3
LK-41	sV	Fluvial	Swr	vert.	W-E			44.3	11.3	0.0	0.0	0.3	0.3	2.3	0.0	0.7	2.7	0.3	0.0	2.3
LK-42	sV	Fluvial	Swr	vert.	W-E			61.7	10.0	0.0	0.0	0.3	0.0	1.0	0.0	0.0	1.0	0.0	0.7	0.7
LK-43	sV	Fluvial	Swr	vert.	W-E			58.7	8.0	0.0	0.0	0.0	0.3	2.0	0.0	0.0	0.3	0.0	0.7	0.0
LK-44	sV	Fluvial	Swr	vert.	N-S			64.7	8.0	0.0	0.0	0.0	1.0	1.0	0.3	0.0	0.7	0.3	0.0	0.0
LK-45	sV	Fluvial	Swr	vert.	N-S			62.7	10.0	0.0	0.0	0.3	0.7	1.0	0.3	0.3	1.7	1.0	0.0	0.3
LK-46	sV	Fluvial	Swr	vert.	N-S			52.3	8.7	0.0	0.0	0.0	0.0	0.7	0.0	0.0	2.0	0.3	0.3	0.3

sV: Vogesensandstein Formation, St: trough cross-bedded sandstone, Swr: wavy to ripple-bedded sandstone, hor.: horizontal, vert.: vertical

Table A.2: (continued).

Sample Name	Accessories	Iron Oxide	Detrital clay	*Authigenic components [%]* Quartz cement	Illite tangential	Illite radial	Illite meshwork	Alkali feldspar cement	Kaolinite undifferentiated	Carbonate pore-filling	Barite cement	Illite replacing alkali feldspar	Carbonate replacing undifferentiated grain	*Porosity [%]* Intergranular porosity	Secondary porosity - Alkali feldspar	Secondary porosity - Rock fragment	Secondary porosity - undifferentiated	Secondary porosity - after nodule dissolution
LK-36	0.0	6.7	0.3	9.7	0.3	1.0	0.0	0.3	0.3	0.0	0.0	1.7	0.0	11.3	1.7	0.0	0.3	0.0
LK-37	0.0	6.7	0.0	9.7	0.3	0.7	0.3	0.7	0.0	0.0	0.0	1.0	0.0	4.0	1.0	0.0	0.3	0.0
LK-38	0.3	8.7	0.0	11.7	2.3	4.3	2.0	2.0	0.0	0.0	0.0	3.3	0.0	2.3	1.3	0.0	0.3	0.0
LK-39	0.0	6.3	0.0	13.3	0.7	1.7	0.0	0.3	0.0	0.0	0.0	1.3	0.0	11.3	0.7	0.0	0.3	0.3
LK-40	0.3	6.7	0.0	2.3	3.0	2.0	0.0	0.3	0.7	0.0	0.7	3.0	0.0	2.7	1.0	0.3	0.0	0.0
LK-41	0.3	19.0	0.3	2.7	3.7	0.7	0.7	1.0	1.0	0.0	0.0	3.0	0.0	0.7	1.3	0.3	0.7	0.0
LK-42	0.7	12.0	0.0	2.3	2.0	2.3	0.3	0.0	0.0	0.0	0.0	1.3	0.0	2.3	1.3	0.0	0.0	0.0
LK-43	0.0	9.0	0.0	4.7	2.3	4.3	0.7	0.3	0.3	0.0	0.0	0.3	0.0	6.0	2.0	0.0	0.0	0.0
LK-44	0.3	11.7	0.0	2.3	1.7	2.7	1.3	0.0	0.7	0.0	0.0	1.3	0.0	1.3	0.3	0.0	0.3	0.0
LK-45	0.0	8.3	0.0	3.0	1.0	2.3	1.0	0.3	0.0	0.0	0.0	1.0	0.0	2.7	1.7	0.0	0.3	0.0
LK-46	0.0	20.0	0.0	1.7	1.7	3.0	1.3	0.3	1.0	1.7	0.0	2.3	0.7	1.3	0.3	0.0	0.0	0.0

Table A.2: (continued).

Sample Name	*Spectral gamma ray SGR*				*Additional sample information*						
	K [%]	U [ppm]	Th [ppm]	SGR [API]	Plug porosity [vol%]	Plug permeability [mD]	Point count porosity incl. microporosity [vol%]	Mean grain size [mm]	Mean grain size [Phi] (Udden, 1914 & Wentworth, 1922)	Grain size class (Udden, 1914 & Wentworth, 1922)	Sorting class (Folk & Ward, 1957)
LK-36	N/A	N/A	N/A	N/A	20.2	451.235	15.5	0.276	1.962	mg	ms
LK-37	N/A	N/A	N/A	N/A	12.7	0.108	6.8	0.304	1.872	mg	ms
LK-38	N/A	N/A	N/A	N/A	16.8	0.518	11.6	0.192	2.508	fg	ms
LK-39	N/A	N/A	N/A	N/A	20.2	281.868	15.0	0.271	1.953	mg	ws
LK-40	N/A	N/A	N/A	N/A	9.3	0.004	9.3	0.283	1.926	mg	ms
LK-41	N/A	N/A	N/A	N/A	11.1	0.004	8.6	0.168	2.758	fg	ms
LK-42	N/A	N/A	N/A	N/A	10.4	0.006	7.4	0.231	2.216	fg	ms
LK-43	N/A	N/A	N/A	N/A	13.9	0.022	13.0	0.302	1.860	mg	ms
LK-44	N/A	N/A	N/A	N/A	10.1	0.005	6.7	0.306	1.840	mg	ms
LK-45	N/A	N/A	N/A	N/A	11.1	0.009	8.0	0.231	2.307	fg	ms
LK-46	N/A	N/A	N/A	N/A	10.0	0.003	7.3	0.231	2.300	fg	ms

v. fg: very fine-grained, fg: fine-grained, mg: medium-grained, ms: moderately sorted, ws: well sorted

Table A.2: (continued).

Sample Name	*Compaction parameters* IGV [vol%]	COPL [vol%]	CEPL [vol%]	*Grain contact types [%]* none	point contact	long contact	concavo-convex contact	sutured contact
LK-36	23.3	28.3	8.5	4	19	41	36	0
LK-37	15.7	34.8	7.6	4	12	43	41	0
LK-38	24.7	27.0	16.3	1	7	24	67	1
LK-39	27.3	24.3	12.1	7	25	34	34	0
LK-40	11.7	37.7	5.6	0	1	4	89	6
LK-41	10.3	38.7	5.9	0	0	6	88	6
LK-42	9.3	39.3	4.2	1	0	5	92	2
LK-43	18.7	32.4	8.6	0	1	20	78	1
LK-44	10.0	38.9	5.3	0	0	7	93	0
LK-45	10.3	38.7	4.7	0	1	8	91	0
LK-46	12.0	37.5	6.7	0	0	11	89	0